T0259583

Engineering Ethics

Engineering Ethics: An Industrial Perspective

Gail D. Baura

AMSTERDAM • BOSTON • HEIDELBERG • LONDON
NEW YORK • OXFORD • PARIS • SAN DIEGO
SAN FRANCISCO • SINGAPORE • SYDNEY • TOKYO

Academic Press is an imprint of Elsevier

Cover photo credits:
Space Shuttle photos reprinted from NASA *Report of the Columbia Accident Investigation Board*, 2003.
San Francisco–Oakland Bay Bridge photo courtesy of U.S. Army Corps of Engineers.
Exxon Valdez oil spill cleanup photo courtesy of the Exxon Valdez Oil Spill Trustee Council.

Elsevier Academic Press
30 Corporate Drive, Suite 400, Burlington, MA 01803, USA
525 B Street, Suite 1900, San Diego, California 92101-4495, USA
84 Theobald's Road, London WC1X 8RR, UK

This book is printed on acid-free paper. ∞

Library of Congress Cataloging-in-Publication Data
Application submitted

British Library Cataloguing in Publication Data
A catalogue record for this book is available from the British Library

ISBN 13: 978-0-12-088531-2
ISBN 10: 0-12-088531-X

For all information on all Elsevier Academic Press publications
visit our Web site at www.books.elsevier.com

Printed and bound in Great Britain by
CPI Antony Rowe, Chippenham and Eastbourne
Transferred to Digital Printing, 2010

Dedication

To Larry Spiro, my *bon vivant*, without whose infinite patience and love this book could not have been written.

To my friends, Bob Ward and Sandy Ng, whose stories inspired this book.

To my brother-in-law, Steve Conklin, who taught me about Qui Tam.

To Dr. David Graham, who is one of my heroes.

To the Anonymous Seven, for their bravery.

The only rational way of educating is to be an example—if one can't help it, a warning example.

—Albert Einstein

Contents

Foreword   xv
Preface   xvii

Part I
An Ethics Foundation

Chapter 1

A Personal Engineering Ethics Threshold

A Real World Example   4
What Is Engineering Ethics?   6
Ethical Theories   7
  Utilitarianism   9
  Duty Ethics   10
  Rights Ethics   11
  Virtue Ethics   12
Engineering Ethics Codes   13
  NSPE Code of Ethics for Engineers   13
  IEEE Code of Ethics   13
  Code Effectiveness   14
Professional Responsibility   14
  Protection of Public Safety   14
  Technical Competence   15
  Timely Communication of Negative & Positive Results
    to Management   15
Ethical Dilemmas   16
  Public Safety & Welfare   16
  Data Integrity & Representation   17
  Trade Secrets & Industrial Espionage   17
  Gift Giving & Bribery   17
  Principle of Informed Consent   17

Conflict of Interest   18
Accountability to Clients & Customers   18
Fair Treatment   18
Determining Your Personal Engineering Ethics Threshold for Action   18
What Is Your Personal Threshold?   19
References   20
Questions for Discussion   20

Chapter 2

Options for Action When an Engineering Ethics
Threshold Is Reached

Departure   23
Whistleblowing   24
The Employee Conscience   25
Employee Protection Legislation   26
Employee Protection Procedures   28
Employee Protection Examples   29
The Observer Conscience   30
Observer Protection Legislation   30
Observer Protection Procedures   31
Observer Protection Examples   32
Conclusion   33
References   33
Questions for Discussion   34

Part II

National Case Studies

Chapter 3

1978: Ford Pinto Recall

The Reported Story   39
The Back Story   39
Automobile Safety   39
Ford Subcompact Car Project   41

Pinto Investigations   45
Pinto Lawsuits   46
Applicable Regulations   47
An Engineering Perspective   49
References   50
Questions for Discussion   51

Chapter 4

1981: Kansas City Hyatt Regency Skywalk Collapse

The Reported Story   53
The Back Story   53
Kansas City Hyatt Regency Hotel Design   53
Hyatt Project Hierarchy   54
Original Box Beam Hanger Rod Design
    and Modifications   55
Atrium Roof Collapse   57
Walkway Investigation   57
Administrative Hearing Actions   58
Applicable Regulations   59
An Engineering Perspective   59
References   59
Questions for Discussion   60

Chapter 5

1986: Challenger Space Shuttle Explosion

The Reported Story   63
The Back Story   63
The Space Shuttle Design   63
Early Problems   67
Launch Delays and Subsequent Launch   68
Presidential Commission Investigation   69
Commission Recommendations   70
Applicable Regulations   71
An Engineering Perspective   71
References   73
Questions for Discussion   74

Chapter 6
1989: Exxon Valdez Oil Spill
The Reported Story   75
The Back Story   75
   The Trans-Alaska Pipeline System   75
   Oil Spill Preparedness   78
   The Last Voyage of the Exxon Valdez   79
   Oil Spill Cleanup   80
   Oil Spill Investigation   81
   Lawsuits   82
Applicable Regulations   82
An Engineering Perspective   85
References   86
Questions for Discussion   87

Chapter 7
1989: San Francisco–Oakland Bay Bridge Earthquake Collapse
The Reported Story   89
The Back Story   89
   Transportation in the Bay Area in the 1920s   89
   Bridge Design and Construction   90
   Railway Retrofit   91
   Loma Prieta Earthquake   93
   New Bridge   94
Applicable Regulations   95
An Engineering Perspective   96
References   97
Questions for Discussion   97

Chapter 8
1994: Bjork-Shiley Heart Valve Defect
The Reported Story   99
The Back Story   99
   Heart Valves   99
   The Bjork-Shiley Heart Valve   100

Bjork-Shiley Valve Complications & Investigations   101
U.S. Government Intervention   104
Other Lawsuits   104
Applicable Regulations   105
An Engineering Perspective   107
References   107
Questions for Discussion   108

Chapter 9

1999: Y2K Software Conversion

The Reported Story   109
The Back Story   109
The Millennium Bug   109
Special Committee on the Year 2000 Technology
    Problem   110
Interim Assessments   110
Day One Preparation   112
Y2K Aftermath   112
Applicable Regulations   114
An Engineering Perspective   116
References   116
Questions for Discussion   117

Chapter 10

2002: Bell Laboratories Scientific Fraud

The Reported Story   119
The Back Story   120
The History of Bell Laboratories   120
Nanotechnology   123
Jan Hendrik Schon's Bell Labs Work   124
Bell Labs Investigation   126
Aftermath   128
Applicable Regulations   128
A Scientific Perspective   130
References   130
Questions for Discussion   131

Chapter 11

2002: Ford Explorer Rollover

The Reported Story   **133**
The Back Story   **133**
    Ford Explorers and Firestone Tires   133
    CAFE Standards and Sport Utility Vehicles   134
    The Ford Bronco II   135
    The Ford Explorer   135
    Government Regulation of SUVs   137
    Ford Explorer Lawsuits   138
    The 2002 Ford Explorer   139
Applicable Regulations   **139**
Business & Professions Code   **140**
An Engineering Perspective   **140**
References   **142**
Questions for Discussion   **143**

Chapter 12

2003: Columbia Space Shuttle Explosion

The Reported Story   **145**
The Back Story   **145**
    External Tank Insulation   146
    Early Problems   147
    Launch Delays and Subsequent Launch   148
    Columbia Accident Investigation Board   148
    Investigation Board Recommendations   149
Applicable Regulations   **150**
An Engineering Perspective   **151**
References   **151**
Questions for Discussion   **152**

Chapter 13

2003: Guidant Ancure Endograft System

The Reported Story   **153**
The Back Story   **154**
    FDA Good Manufacturing Practices   154
    Medical Device Background   154

      Clinical Performance After FDA
         Approval   156
      Device Investigations   157
      Settlement   158
Applicable Regulations   **158**
An Engineering Perspective   **160**
References   **160**
Questions for Discussion   **161**

Chapter 14

2003: Northeast Blackout

The Reported Story   **163**
The Back Story   **163**
      The North American Interconnection   163
      Grid Reliability   165
      The Grid in Ohio   166
      The Initiation of the Blackout   167
      Task Force Investigation and
         Recommendations   168
Applicable Regulations   **171**
An Engineering Perspective   **173**
References   **173**
Questions for Discussion   **174**

Chapter 15

2004: Indian Ocean Tsunami

The Reported Story   **177**
The Back Story   **177**
      The Sumatra-Andaman
         Earthquake   177
      The Indian Ocean Tsunami   179
      Tsunami Aftermath   180
      Tsunami Warning Systems   181
Applicable Regulations   **182**
An Engineering Perspective   **182**
References   **183**
Questions for Discussion   **184**

Part III
Individual Case Studies

Chapter 16
Anonymous Industrial Engineering Ethics Cases

Case 1: Biomedical Engineer   187
Case 2: Mechanical Engineer   189
Case 3: Electrical Engineer   191
Case 4: Geologic Engineer   192
Case 5: Biomedical Engineer   194
Case 6: Electrical Engineer   196
Case 7: Mechanical Engineer   197
Case 8: Biomedical Engineer   198
Case 9: Computer Engineer   200
Case 10: Electrical Engineer   201

Appendix

National Society of Professional Engineers (NSPE) Code of Ethics
    for Engineers   203
Institute of Electrical and Electronic Engineers (IEEE)
    Code of Ethics   210
SARBANES-OXLEY ACT OF 2002   211

Index   215

A Request from the Author   219

About the Author   220

Foreword

by
Dennis A. Gioia
Professor of Organizational Behavior
Smeal College of Business
Penn State University

Like many of you perusing this book, I was once a budding young engineer, anticipating a career working on some interesting technical problems for some exciting *Fortune 500* company. I loved engineering. I loved the whole idea of it—from the intense puzzling over some difficult technical challenge, to the intellectual high of discovering the "elegant solution," to the implementation of that solution in practice. Science and engineering were full of elegant solutions just waiting for me to find them, and I thanked my lucky stars that I was not going to be dealing with the messiness of ambiguous decisions and people problems associated with the less technically accomplished students' careers. Engineering had a purity and an idealism that were very appealing to me, as I suspect they are to you.

My first engineering job was with Boeing Aerospace at Cape Kennedy, working on the Apollo/Saturn lunar program as an apprentice. It was the embodiment of every engineering ideal I had ever imagined. That program was *pure* science and engineering, with the loftiest conceivable goal—to put people on the moon and return them safely to earth—but it was so pure in purpose that it left me unprepared for life in the "real world" of business. You probably can imagine my surprise when I joined Ford Motor Company and found a realm of complexity and ambiguity requiring a myriad of judgment calls—some of them involving competing values and murky ethical choices. Sure enough, I found myself working on quite a number of interesting engineering problems, but now many of the solutions to those problems were colored by legal issues, social issues, and moral issues. Some of them, such as the problem with Pinto fuel tank integrity during rear-end collisions, even involved choices with life-and-death implications. Heavens. My engineering training hadn't prepared me for anything like this!

My experience in the automotive industry, however, turned out to be closer to the norm than the exception. These days, most engineering decisions are couched within some larger domain involving social, political, and ethical choices. Engineering is no longer a pure discipline (if it ever was). Engineers can no longer practice their skills independent of these larger domains but rather must consider them as a constituent part. For that reason, in the modern era, it is no longer acceptable to become an engineer without an appreciation of the ethical context within which we serve society, usually through our work in industry. And that's why it is important for you to read this book and consider with care the instructive cases that harbor lessons for you as you prepare for your now more complicated career as a responsible engineer.

Preface

While I was finishing my MS degree and conducting interviews for my first job in industry in 1986, the Challenger space shuttle exploded. As the details came out that Chief O-ring engineer Roger Boisjoly had recommended the launch be postponed, I found it incomprehensible that expert engineering opinions would be ignored. Now, with 16 years of industrial experience, I realize how frequent this situation has become. As former Treasury Secretary Paul O'Neill noted in *The Price of Loyalty*, "there had been an ongoing shift, across nearly two decades, of what is acceptable conduct for a corporation . . . . There were tens of thousands of companies in America . . . that were operating with virtually no proactive standard to compel probity" (Suskind, 2004).

I decided to write this book after a difficult 12-month period in engineering ethics. On September 25, 2002, scientific fraud at Bell Laboratories was exposed. While truly a question of scientific, rather than engineering, ethics, Bell Labs was the first company for which I worked as an engineer. I could not imagine how a former paragon of ethics had fallen. On February 1, 2003, the Columbia space shuttle exploded. Although initial reactions included suspicions of terrorist activity, I suspected that when the facts came out, it would be proved again that expert engineering opinions were ignored. On June 12, 2003, Guidant Corporation agreed to the largest payout ever, $92.4 million, for violating the Food and Drug Administration's medical device report requirements for its Ancure Endograft System. Finally, on August 14, 2003, the Northeast blackout occurred.

The thesis of this textbook is that within the course of their industrial careers, many new engineering graduates will be exposed to serious ethics violations. Thirteen detailed case studies, including the examples above, are given of situations in which an engineer (or in the case of Bell Labs, a scientist) warned his superiors of a potentially grave situation but was ignored. Unlike case studies in other engineering ethic texts, these cases are not written in narrative form. Rather, each is presented in the following format: 1) the reported (newspaper) story, 2) the back story,

3) applicable regulations, 4) an engineering perspective (warnings before the event), and 5) questions for discussion.

To complement these case studies, a discussion of personal responsibility and how an engineer sets his or her personal engineering ethics threshold is presented in Chapter 1. This chapter includes descriptions of some of the major ethical theories (Utilitarianism, Duty Ethics, Rights Ethics, and Virtue Ethics) and references to engineering ethics codes. In Chapter 2, options for engineering actions when this personal threshold is reached are detailed. In Chapter 16, actual anonymous industrial cases of engineering ethics are presented that include an abbreviated description of each situation and how each engineer responded. The engineers who shared these personal experiences want students to be prepared for ethical situations before they encounter them in industry.

This text may be used within an Introduction to Engineering or Senior Design course if it is decided that engineering ethics be taught within another course to meet the ABET ethics requirement. Alternatively, this text may be used as an adjunct to any of the new engineering ethics textbooks because most do not provide extensive case studies or a concentrated industrial perspective. Although two current textbooks provide 31 and 57 case studies, these case studies are not as comprehensively detailed as the case studies in this book. Almost all of the other textbook authors are professors without industrial engineering experience. I believe that incorporating an industrial perspective is important in an engineering ethics course.

Ultimately, this text does not provide solutions. I believe it is necessary that students discuss engineering ethics in school and, during their first year in industry, come to an understanding of the industrial culture in which they function. With an engineering ethics foundation, I hope each will then be able to choose his or her personal engineering ethics threshold and determine a suitable course of action when this threshold is reached.

The impetus for this book was an engineering ethics discussion I had with Drs. John Enderle and Jerry Jakabowski during an ABET visit in the fall of 2003. Reflecting on this discussion a few months later over Christmas vacation, I noticed the similarities with discussions I had with my friends, Bob Ward and Dr. Sandy Ng. As I began to discuss engineering ethics with my very patient family, my brother-in-law, Steve Conklin, taught me about Qui Tam.

I would like to thank my dear friends, Fred Bacher, Simon Finburgh, and James Grove, for encouraging and editing this manuscript. Dr. Dan Porte, Jr., provided helpful reviews of foundational Chapters 1 and 2. My editors Joel Stein and Shoshanna Grossman supplied valuable feedback during the writing process. My friend Paula Mason acted as the spiritual mentor for this project. The 10 anonymous engineers who volunteered their personal experiences for Chapter 16 offer a unique glimpse into the engineering work environment.

My husband, Larry, continues to be supportive of my obsessive endeavors. In our 18 years together, he has opened my mind up to so much more than system theory. And I am so, so sorry that my first book wasn't my last book.

My Ph.D. advisors, Drs. David Foster and Dan Porte, Jr., taught ethics by example. I wish this type of mentoring were still sufficient to ensure an ethical work environment.

I welcome comments to this text at www.gailbaura.com.

Gail D. Baura

Suskind, R., *The Price of Loyalty: George W. Bush, the White House, and the Education of Paul O'Neill*. New York: Simon & Schuster, 2004, 225.

Part I

An Ethics Foundation

As we practice engineering, our decisions are generally guided by the project management variables of cost, schedule, and quality. If you change one of these variables, the ones remaining will also be changed. But our decisions are also guided by our moral values; that is, our concern and respect for others. The framework of our ethical decisions is based on ethical theories we have learned and ethical behavior we have observed during our lives. It is advocated in Chapter 1 that engineers use this framework, as well as three professional responsibilities, to guide their professional behavior. These three responsibilities—concern for public safety, technical competence, and timely communication of positive and negative results to management—are grounded in respecting others and keeping them safe.

During their careers, many engineers will become involved in unethical situations they cannot control. In Chapter 1, possible ethical dilemmas that may occur are discussed. It is up to each engineer to determine his or her personal engineering ethics threshold for action; that is, which ethical dilemmas may occur at a workplace before the engineer is forced to an extreme action of leaving the company or fighting for change. In Chapter 2, we discuss how to perform these extreme actions.

Part I

An Ethics Foundation

Chapter 1

A Personal Engineering Ethics Threshold

A warning is in order as you begin to read this book: *this textbook is different.* Unlike other engineering ethics texts, this one is written from a personal perspective by an engineer who currently works in industry. Over the course of two decades, I have witnessed a decline in business ethics, which culminated with the Enron and WorldCom scandals of 2001–2002. My own anger and disbelief stems from the Bell Laboratories nanotechnology fraud of September 2002, during which the fundamental nanotechnology results of one scientist were found to be completely fabricated, leading to retractions of articles in the journals *Science* and *Nature.* Having worked at AT&T Bell Labs in the 1980s, when it was known for its high standards of excellence, I could not understand how its operating procedures could have plummeted (Baura, 2005). Of course, this was before I discovered that Lucent, which now owned Bell Labs, improperly reported $1.148 billion in revenue and $470 million in pretax 2000 income, causing the Securities and Exchange Commission to fine Lucent $25 million (Young and Berman, 2004); was still in recovery from the telecom crash; and lost $28 billion over a 24-month period, from 2001 to 2003 (Berman, 2003).

I view the Sarbanes-Oxley (SOX) Corporate Reform Act of 2002 as a positive step. Although not perfect, it goes a long way toward making public corporations accountable for their behavior. SOX has created an environment in which corporations have realized the importance of

3

instituting ethics policies and codes of conduct to address issues related to unethical or illegal conduct. Although none of the other texts mention SOX, I believe it will enable engineers to conduct their jobs in an ethical manner.

A REAL WORLD EXAMPLE

We begin our discussion of how to determine our personal threshold for action by examining the real world events that geodesic engineer Jack Spadaro encountered. Spadaro was second in command of a team selected by the Mine Safety and Health Administration (MSHA) to investigate a coal slurry spill that occurred on October 11, 2000. *Coal slurry* refers to the wastewater and impurities that result from coal washing and processing. An embankment made of coarse coal refuse acts as a dam to contain the slurry at a mining site. As sediment settles out of this mixture, filling the pond, wastewater is recycled back into the coal washing process. The slurry pit remains after mining operations cease.

During this disaster, a slurry pit in Inez, Kentucky, owned by a subsidiary of Massey Energy, burst into subsurface mine shafts, flooding downstream communities. This 300-million gallon spill was the largest in American history. According to the Environmental Protection Agency, it was the greatest environmental catastrophe in the history of the eastern United States. Thick, black, lava-like toxic sludge containing 60 poisonous chemicals choked and sterilized 100 miles of rivers and creeks and poisoned the drinking water in 17 communities.

Spadaro, the former superintendent of the National Mine Health and Safety Academy, where MSHA trains its engineers, is nationally recognized for his slurry spill expertise, having spent 30 years studying slurry dam failure and prevention. In the course of the team's investigation, it was discovered that mitigation of a spill at the same site in 1994 had been misrepresented to the government (Kennedy, 2004). Mining officials had stated that a solid coal barrier at least 70 to 80 feet wide between the mine workings and the bottom of the reservoir existed, when in fact the barrier was less than 20 feet. An engineer at the Massey subsidiary, Martin County Coal, admitted he and the company knew another spill was inevitable (Simon, 2004). Martin County Coal has stated that the slurry spill was accidental.

Massey Energy is the fifth largest U.S. mining company, and a large contributor to the Republican party. When the Bush administration took over the White House in 2001, the MHSA team was given a smaller scope of investigation, and was asked to complete its investigation in a few weeks. On the day of Bush's inauguration, Tony Oppegard, Spadaro's boss whom the team regarded as a strong leader with unquestioned integrity, was fired. Although the team had originally intended to cite Massey for eight violations of criminal negligence, this was ultimately reduced to only two violations. One of these was later thrown out by the administration judge, and Massey was fined $5,600 in respect of the remaining violation (Kennedy, 2004). However, Massey also paid $3.25 million in penalties to the state of Kentucky, $225,000 to the Department of Fish and Wildlife Resources, $46 million for clean-up, and unspecified amounts to more than 400 local residents in out-of-court settlements (Alford, 2005).

Spadaro refused to sign the final report documenting this investigation. As Spadaro later told Robert F. Kennedy, Jr., senior attorney for the Natural Resources Defense Council, "I've been regulating mining since 1966, and this is the most lawless administration I've encountered. They have no regard for protecting miners or the people in mining communities. They are without scruples. I know that Massey Energy influenced Bush appointees to alter the outcome of our report. The corruption and lawlessness goes right to the top" (Kennedy, 2004). Specifically, Spadaro alleged that Kentucky Senator Mitch McConnell, whose wife is Department of Labor Secretary Elaine Chao, tried to protect Massey because it is a major campaign contributor. MSHA is part of the Department of Labor. When the New York Times attempted to investigate Spadaro's allegation, Mr. McConnell declined to comment (Dao, 2003).

Government agents later raided Spadaro's office, searched his papers, and changed the locks. Spadaro was not allowed to return to work. When the federal government's independent Office of Special Counsel began an investigation in February 2004 to determine if Spadaro was being disciplined as a whistleblower, MSHA demoted him 1 week later and reassigned him to a job in Pittsburgh. The job included a $35,000 pay cut (Simon, 2004). Spadaro appealed the demotion and transfer but settled with MSHA in October 2004. Per the terms of the settlement, he agreed to drop his appeal and MSHA discontinued any actions against him. MSHA also restored Spadaro's previous pay grade, which if not done would have reduced Spadaro's retirement benefits. Immediately after the settlement, Spadaro retired (Associated Press, 2004).

Jack Spadaro paid a huge price for questioning the actions of his agency. Throughout his career, he practiced the professional responsibilities of public safety protection, technical competence, and results reporting to management. However, in return for trying to protect miners and their communities after the Inez coal slurry spill, Spadaro was initially not allowed to work and was later demoted and transferred. Only after appealing these actions and having his case publicized by the *60 Minutes* television program were these actions reversed.

How did Spadaro arrive at his decisions? Let us examine engineering ethics tools that enable this decision-making process. After defining engineering ethics, we discuss the classic engineering ethics tools of ethical theories and ethics codes. I then highlight three professional responsibilities that are a subset of the ethics codes.

WHAT IS ENGINEERING ETHICS?

As stated by Schinzinger and Martin, "Engineering ethics . . . is the study of the moral values, issues, and decisions involved in engineering practice" (Schinzinger and Martin, 2000). Morality encompasses the first-order beliefs and practices about good and evil by which we guide our behavior. Ethics is the second-order, reflective consideration of our moral beliefs and practices (Hinman, 2003).

In the course of practicing engineering, an engineer solves problems. But because there is no perfect solution, any implemented solution inevitably creates a new problem. The new problem may be small, such as developing a software algorithm that fulfills customer expectations but requires so much software program memory that only one more software upgrade is possible using the current hardware. Or the new problem may be large, such as saving program memory by omitting the first two digits of the year during the 1900s, which caused the Y2K scare at the turn of this century.

As we practice engineering, our decisions are generally guided by the project management variables of cost, schedule, and quality. If you change one of these variables, the ones remaining will also be changed. But our decisions are also guided by our moral values; that is, our concern and respect for others. Further, local, state, and federal laws may influence our behavior. In the course of this chapter, you will learn about three professional responsibilities I believe every engineer should always follow, which are but a subset of responsibilities advocated by engineering societies. These three responsibilities—concern for public safety, technical competence, and timely communication of positive and negative results to management—are grounded in respecting others and keeping them safe. As engineers, we are involved in so many projects that touch people's lives. It is important that we protect the consumers of our technologies.

My industrial colleagues and I are grateful that the Accreditation Board of Engineering Technology (ABET), which accredits U.S. undergraduate engineering programs, mandated that ethics topics be incorporated into undergraduate engineering curricula. Similarly, the Association to Advance Collegiate Schools of Business, which accredits Masters of Business Administration (MBA) programs, increased the emphasis on ethics in its 2004 curriculum learning standard (AACSB, 2004). We wish that this type of study had been available to us in the classroom. Some of my colleagues felt strongly enough that they contributed their personal engineering dilemmas as the anonymous case studies found in Chapter 16. We believe that engineering ethics should be taught in school to provide engineers entering industry with a foundation for ethical behavior. The corporate culture is very powerful and can sway a young engineer's thinking.

For example, Dennis Gioia was promoted to field recall coordinator at Ford Motor Company in 1973, only 2 years out of school (Bachelor of Science [BS] in Engineering Science, MBA). Part of his new position involved making initial recommendations about possible future recalls. Although he received reports of Pinto fires after low-speed rear-end collisions, Gioia did not recommend a Pinto recall. He does remember, however, "being disquieted by a field report accompanied by graphic, detailed photos of the remains of a burned-out Pinto in which several people had died" (Gioia, 1992). Writing about the Ford Pinto experience 19 years later, Gioia stated that his "own schematized (scripted) knowledge influenced me to perceive recall issues in terms of the prevailing decision environment and to unconsciously overlook key features of the Pinto case, mainly because they did not fit an existing script" (Gioia, 1992).

By discussing the foundation for engineering ethics and reviewing national headlines and personalized case studies, my colleagues and I hope that you will be better prepared to enter the industrial environment. We will consciously reflect on our moral beliefs within the context of corporate situations, extending and refining these beliefs. By practicing ethical analysis, we will strengthen our ability to conduct it. It is important to integrate your professional life with personal convictions in order to maintain your moral integrity.

ETHICAL THEORIES

We begin our discussion of an engineering ethics foundation by considering some of the classic ethical theories: utilitarianism, duty ethics, rights ethics, and virtue ethics. Realizing that a thorough treatment of each theory would result in four textbooks, we will restrict ourselves to highlights of each theory. Most of our discussion is based on the ethics textbook by

Lawrence Hinman (2003). Using each of these theories, we will determine how an engineer would respond to the following fictional situation:

Company X has the organizational structure of four vice presidents (VPs) reporting to a chief executive officer (CEO). Together, the CEO and VPs make up the Executive Committee. The VP of Engineering, Mr. Early Retirement, manages the design and testing of new products. He has two direct reports: (1) the hardware manager, Mr. Concerned, who designs and tests electronic circuits with his engineering staff, and (2) the software manager, Mr. Incompetent, who designs and tests software with his engineering staff. The VP of Operations is responsible for managing the manufacture and delivery of products.

Mr. Concerned was hired $1\frac{1}{2}$ years ago to replace the hardware manager fired for poor circuit design, which directly resulted in a product recall. A secondary cause of the recall was software that included incorrect parameter calculations was released by Mr. Incompetent. The new project in engineering is not going well. Because Mr. Early Retirement is very lax in his management style, the new project is several months behind its milestone of transferring a completed design to Operations for manufacture.

Two months after the original transfer date passed, Mr. Early Retirement sent out an e-mail announcing to the Executive Committee, Engineering, Operations, and Marketing that the product had been transferred to Operations. The announcement initiated the Executive Committee's finalization of a plan for product market release and sales. But the announcement was not true, because one of the circuit boards had a problem that had not been solved, and several software lockups were known to exist. *Lockup* refers to a product stalling after some interaction with the user, which can only be fixed by powering the product off and then on. Two weeks later, Mr. Incompetent decided, in his words, to "gut the state machine" to prevent software lockups. The state machine is the "brains" of the product, which specifies the sequences of states that an object or an interaction goes through during its life in response to events, together with its responses and actions. The state machine is typically designed at the beginning, not the end, of a project. In a meeting, Mr. Concerned observed that Mr. Early Retirement and Mr. Incompetent, who were very good friends, were not concerned with repercussions of a state machine overhaul.

Mr. Concerned believes a product recall may occur because of software if adequate software testing and subsequent software revisions are not completed before product shipment. Mr. Early Retirement and Mr. Incompetent disagree.

Should Mr. Concerned tell the CEO, who is Mr. Early Retirement's direct supervisor, that (1) the e-mail announcement of transfer was false, (2) the product is still not ready to transfer, and (3) a product recall may occur unless adequate testing is completed?

UTILITARIANISM

According to utilitarianism, the morality of an action is determined solely through an assessment of its consequences. Originally, utility was defined by Jeremy Bentham in terms of actions that maximized pleasure and minimized pain. However, this "pig's philosophy" was later reformulated by Bentham's godson, John Stuart Mill, to maximize happiness and minimize suffering. Optimizing happiness, rather than pleasure, seems a better choice, as happiness is related more to the mind than the body, is of longer duration, and may encompass both pleasure and pain (e.g., childbirth).

Using a relative scale, potential units of happiness, called hedons, may be compared with potential units of suffering, called dolors, to determine if an action should be pursued. It is the relationship of relative happiness to suffering that the utilitarian seeks to capture in assigning numerical values to various consequences. For example, voting to reduce medical benefits for the elderly may result in 10 hedons each for 100 million people and 200 dolors each for 20 million people and 3 dolors each for 100 million people, with an overall utility of 3.3 billion dolors. In contrast, keeping these benefits the same may result in 20 hedons each for 20 million people and 3 dolors for 100 million people, with a total overall utility of 100 million hedons. Alternatively, increasing these benefits may result in 90 hedons apiece for 20 million people and 20 dolors for 100 million people, with a total overall utility of 200 million dolors. Comparing these three alternatives from a utilitarian perspective, we would be obligated to vote for keeping benefits the same, because the other courses of action have a lesser overall utility.

Utilitarianism has several limitations. First, it is difficult to weigh matters of life and death by weighing happiness against suffering. Second, utilitarianism is unable to distinguish between morally justified and morally unjustified emotions. For example, a thief may derive great happiness from stealing money from others. Finally, utilitarians may not give special weight to the fact that certain consequences may affect them personally. Even if the utilitarian legislator will suffer personally without an increase in benefits, he or she is still required to vote against an increase if that increase would yield a lesser total utility than the alternatives.

Utilitarian Decision

In our fictional situation, telling the CEO that Mr. Early Retirement is wrong may personally cost Mr. Concerned 100,000 dolors in wrath from his supervisor, as this admission would strain the relationship between Mr. Concerned and Mr. Early Retirement. However, suppressing this information from the CEO may cost the company 20 million dolors to recall the product and pacify customers. Using utilitarianism, Mr. Concerned is obligated to tell the CEO that the product is not ready to transfer to manufacturing.

DUTY ETHICS

According to duty ethics, which was first postulated by Immanuel Kant, an action is moral if it is conducted for the sake of duty, if its maxim can be willed as a universal law, and if it is a respectful way to treat humanity. By acting out of duty, a person acts out of a concern for what is morally right, not out of some self-serving motive. Kant defines a maxim as the subjective rule a person has in mind while performing an action. If a maxim can be consistently adopted as a guide for action, and is thus universal, then it is the moral action of several alternatives. Respect, for Kant, applies to the way a person treats others. In his writings, Kant instructs us to treat others as an end, but never simply as a means to an end.

For a maxim to be considered universal, it must be consistent, impartial, and fair. Consequences should not be considered, in order to consistently will that everyone adopt this maxim. A law should apply equally to all. Exceptions are allowed, as long as they are universal exceptions. For example, if you begin to speed when a friend seems to have had a heart attack in your vehicle as you rush to the nearest hospital, this is an exception to a traffic law. However, this exception would presumably be granted to anyone in an emergency situation and is thus still universal.

Duty ethics has several limitations. First, the exclusive emphasis on duty as the sole motive of moral action may lead to moral alienation. A person may help others out of duty but may not care about helping them. Closely related to this neglect of moral integration of reason with emotion is Kant's exclusion of the emotions from any positive role in a moral life. Kant believed emotions threatened to overwhelm our commitment to good and are external to identity. Finally, duty ethics does not consider moral consequences. Kant believed that if the moral worth of our actions depended on consequences, it would make morality a matter of chance. Of course, consequences do count and cannot be ignored.

Duty Ethics Decision

In our fictional situation, Mr. Concerned is bound by duty to protect the quality of the products his company produces. The maxim of producing high-quality products that are not recalled may be considered universal. Company customers deserve to receive products of high quality that function as advertised. Using duty ethics, Mr. Concerned is obligated to tell the CEO that the product is not ready to transfer to manufacturing. The consequences of this action are immaterial.

RIGHTS ETHICS

According to rights ethics, the morality of an action is determined by the right, or permission to act, of a rights holder and the imposed duty of a rights observer when this holder and observer interact. If a duty is negative, the observer refrains from interfering with the rights holder's exercise of the right. If the duty is positive, the observer takes positive steps to ensure the right is respected. For example, for the negative right to free speech, I may say (within certain limits) whatever I want, and other people are obligated (within certain limits) to not interfere with my speech. For the positive right to basic health care, the state is obligated to provide such care, even if I am unable to pay for it. For our discussion, we will assume that rights are primary and that they override all other types of considerations.

Rights that belong to people simply by virtue of their nature are known as natural rights. When the founders of the United States stated in the Declaration of Independence that certain rights are inalienable, they were referring to our natural rights. Natural rights are established by the appeal to self-evidence, the appeal to a divine sanction or guarantee, the appeal to a natural law, and the appeal to human nature. *Self-evidence* refers to the obviousness of a right that should not be questioned. A divine foundation for human rights offers the strongest imaginable basis for claims of natural rights, because there is no stronger power imaginable than God to guarantee these rights. Because the natural order was created by God, the natural is necessarily good and people are entitled to whatever fulfills the natural order. Properties that are distinctively human, such as our ability to reason, are our rights in the natural world.

Rights ethics has several limitations. First, atheists, because they do not believe in God, will not be convinced to take human rights more seriously because these rights are alleged to be founded in God's will. Second, many philosophers maintain that rights are secondary to, and derivative of, other moral considerations. Within duty ethics, rights are often seen as simply correlatives of duties. Within utilitarianism, the existence and enforcement of rights is seen as dependent on considerations of utility. Finally, to see the world exclusively in terms of rights stresses individualism at the expense of community. An exclusive emphasis on rights has a distorting effect on our vision of the moral life because it fails to see the bonds that hold humans together.

Rights Ethics Decision

In our fictional situation, Mr. Concerned has the right to tell the CEO that Mr. Early Retirement is wrong about the product, much as Mr. Early Retirement had the right to announce product completion by e-mail. The concept of a product recall is immaterial to this discussion. Using rights ethics, Mr. Concerned is obligated to tell the CEO that the product is not ready to transfer to manufacturing.

VIRTUE ETHICS

According to virtue ethics, morality is not related to action, but to virtue. Virtue, as defined by Aristotle, is a habit of the soul, involving both feeling and action, to seek the mean in all things relative to us. Here, *the soul* refers to a person's fundamental character, and *the mean* refers to that middle ground between the two extremes of excess and deficiency. Virtue leads to happiness or human flourishing. People flourish politically through participation in the common life of the city-state and contemplatively through a withdrawal from the world of everyday affairs.

Virtues are those strengths of character that promote human flourishing, with human flourishing defined in terms of reasoning or thinking. Perseverance is the ability to act in the face of a difficult and lengthy task. Courage is the ability to act in the face of one's fears. Compassion is the ability to respond to others' suffering in a caring way that seeks to alleviate that suffering or to comfort those who are experiencing it. Self-love is the ability to do whatever promotes your genuine flourishing. When we apply a particular virtue to a particular situation in light of an overall conception of the good life, this is known as practical wisdom.

Virtue ethics has several limitations. First, Aristotle looks for the highest, rather than lowest, common denominator, and considers reason the only character that makes humans unique. By overemphasizing reason, the positive role of emotions and feelings in moral life is neglected. Second, his ethics are for the ruling class only, because much time was to be spent in leisurely contemplation. Fundamentally, virtue ethics fails to tell us how to act because it emphasizes good character over action.

Virtue Ethics Decision

In our fictional situation, Mr. Concerned seeks the mean between Mr. Early Retirement's underestimation of the danger of a recall and overestimation of this danger. Mr. Concerned's courage requires that he tell the CEO that Mr. Early Retirement is wrong. Mr. Concerned's compassion requires that he consider how others will suffer if they use faulty equipment. Using virtue ethics, Mr. Concerned is obligated to tell the CEO that the product is not ready to transfer to manufacturing.

ENGINEERING ETHICS CODES

In the previous section, we reviewed four classic ethical theories and demonstrated how these theories could influence our engineering decision process. However, we are not suggesting that the direct influence of Kant, Aristotle, or another philosopher impacts our daily decisions. Rather, these theories and others we have learned provide a framework for our ethical decisions.

Other sources for our decision framework are the ethics codes of various engineering societies. The codes of the National Society of Professional Engineers (NSPE) and the International Society of Electrical and Electronic Engineers (IEEE) are provided in the Appendix of this text as examples. These and other professional codes establish shared minimum standards and provide guidance and support for responsible engineers.

NSPE CODE OF ETHICS FOR ENGINEERS

NSPE is the only engineering society that represents engineers across all disciplines. Its original code of ethics was approved in 1946. The current code is fairly comprehensive, and details rules of practice as well as professional obligations. Public safety, technical competence, accurate data, avoidance of conflict of interest and other improprieties, professional behavior based on integrity, and professional development are emphasized in this code.

IEEE CODE OF ETHICS

In contrast, the IEEE code is much shorter and more general. The roots of this code date back to the American Institute of Electrical Engineers code adopted in 1912. The current code provides guidelines to protect both engineers and the public. Many of the features of older ethics codes pertaining to topics such as professional courtesy and business ethics for consultants were purposely removed.

CODE EFFECTIVENESS

In some cases, these codes may serve as the formal basis for investigating unethical behavior. In 2001 the IEEE Society on Social Implications of Technology (SSIT) investigated Salvador Castro's ethics case and awarded him the Carl Barus Award for Outstanding Service in the Public Interest. While working as a medical electronics engineer at Air-Shields Inc., Castro discovered a serious design flaw in one of the company's infant incubators. The flaw could be easily and inexpensively fixed, preventing the possibility of infant death. However, Air-Shields chose not to correct the problem. When Castro threatened to tell the Food and Drug Administration, he was fired. IEEE SSIT investigated this case and promised to file an amicus curiae brief on Castro's behalf as Castro's wrongful termination case went to trial (Kumagai, 2004). The preparation and filing of an amicus curiae brief in support of an IEEE member who has upheld the IEEE Code of Ethics is an optional IEEE procedure. This activity exemplifies the support of professional societies for their members' ethics cases, as well as their powerlessness to enforce ethical behavior by their members' employers.

PROFESSIONAL RESPONSIBILITY

With this framework of ethical theories and professional ethics codes as our backdrop, let us discuss the essential requirements for professional responsibility.

PROTECTION OF PUBLIC SAFETY

Engineering projects may directly impact public safety. Whether engineers build bridges or implantable medical devices, the final users of these technologies accept the risks associated with these technologies. Engineers are obliged to inform their supervisors of project risks so that these risks can be communicated to the public if not mitigated in the design.

Designing absolutely safe technologies is impossible, as entirely risk-free activities and products do not exist and no degree of safety satisfies all individuals or groups under all conditions (Schinzinger, 2000). However, it is possible to attain safety through design. Safety through design is defined as "the integration of hazard analysis and risk assessment methods early in the design and engineering stages and the taking of the actions necessary so that the risks of injury or damage are at an acceptable level." This concept encompasses facilities, hardware, equipment, products, tooling, materials, energy controls, layout, and configuration.

A safe design begins with investigation of hazard avoidance, elimination, or control. As design commences, quality standards such as ISO 9000 are adopted. Engineers should consider safety in their design decisions for planned and unplanned maintenance that could affect maintainability and serviceability. They should minimize the probability of failures of equipment and ergonomic risks factors (Christensen, 1999).

TECHNICAL COMPETENCE

Because engineers are required to accomplish tasks demanding specific ability and knowledge, they must be technically competent and conduct themselves competently. When a manager assigns a project task to an engineer, the manager assumes that this task can be completed with high quality in a timely manner. If this task is completed shoddily, or even worse, if the task is incomplete, the entire project is put at risk. In its extreme, this incompetence may endanger public safety, as exemplified by the Kansas City Hyatt Regency skywalk collapse (see Chapter 4). A project engineer at the structural engineering firm working with architects to build this hotel approved a change to the walkway suspension that would cut costs. However, this engineer and his superiors never verified that the modified design was adequate to support reasonable loads nor that it conformed to the Kansas City building code. One hundred fourteen people died when two walkways collapsed shortly after the hotel opened (Stuart, 1981). As Stephen Unger noted in *Controlling Technology,* "The results of incompetence and of malice are often indistinguishable" (Unger, 1994).

When a new engineer has little practical experience, a task may be assigned for which the engineer is not qualified. Rather than hiding this from the engineering manager, who will probably notice soon enough, it is recommended that the engineer admit this up front to the manager. Such responsible behavior will be recognized, and the engineer may receive a mentor for this task or be instructed to take a class. Even if the task is reassigned to another engineer, the new engineer will have acted responsibly and preserved public safety.

TIMELY COMMUNICATION OF NEGATIVE & POSITIVE RESULTS TO MANAGEMENT

During the course of a project, an engineer continuously tests his or her hypotheses and initial designs. An engineer may also test that requirement specifications have been met. These analyses act to move a project task forward. However, depending on a result's relationship to the rest of the project, a manager could interpret these results as detrimental to meeting project milestones. Regardless of possible interpretation, it is an engineer's responsibility to keep the engineering manager informed in a timely manner of negative and positive results. These results should be presented as accurately as possible, with rational discussion of possible consequences. The manager is then responsible for acting on these results.

At least one of the space shuttle disasters could have been averted had managers listened to their engineers (see Chapters 5 and 12). The day before the Challenger exploded in 1986, Morton Thiokol engineers Roger Boisjoly and Arnie Thompson, who had contributed to the solid-propellant booster design, presented their hypothesis and evidence that the forecasted cold weather for the launch would increase problems of joint rotation and joint sealing by the O-rings. Unfortunately, NASA and Morton Thiokol managers chose to ignore the warnings given during the engineers' hour-long presentation and did not postpone the launch. When the O-rings did not seal properly the following day, hot gases escaped from the right solid booster, burning through the external tank. This ignited the liquid propellant, causing the Challenger to explode. Six astronauts and school teacher Christa McAuliffe were killed (World Spaceflight News, 2000).

ETHICAL DILEMMAS

Now that we have a full framework for making ethical decisions, let us discuss the types of ethical dilemmas engineers encounter at work. Please note that in some textbooks, job choice is also considered an ethical dilemma. Job choice may involve ethical decisions, such as whether to work for a military/defense contractor or for a company with a poor environmental record. However, it is not listed here because this dilemma generally occurs before starting a job, not during a job.

PUBLIC SAFETY & WELFARE

As discussed previously, engineering projects may directly impact public safety. Engineers are obliged to inform their supervisors of project risks so that these risks can be communicated to the public. They should attain safety through conscientious design.

DATA INTEGRITY & REPRESENTATION

High-quality engineering analysis starts with careful acquisition of engineering data. Misrepresentation of these data or their subsequent data analysis may disrupt a project. Misrepresentation may take the form of fabrication (inventing data or results), falsification (manipulation of data or results), or plagiarism (appropriation of another's results without proper credit). In the extreme example of Jan Hendrik Schon at Lucent Bell Laboratories (see Chapter 10), fabrication of nanotechnology results caused tens of millions of dollars, including funding from the U.S. Department of Energy, to be wasted. It was estimated that 100 laboratories in the United States and around the world were working on Schon's results by 2002 but could not duplicate them (Cassuto, 2002).

TRADE SECRETS & INDUSTRIAL ESPIONAGE

A trade secret is proprietary company intellectual property that has not been patented. Typically, a new employee signs a confidentiality agreement on the first workday that he or she will not disclose these trade secrets to others, even after leaving for another employer. Industrial espionage may occur when these trade secrets are publicized without consent.

GIFT GIVING & BRIBERY

The acceptance of a gift from a vendor or the offering of a gift to a customer to secure business has the potential to be perceived as a bribe. Company policy should be followed in accepting or giving gifts. Any conflict of interest or appearance of impropriety should be avoided.

PRINCIPLE OF INFORMED CONSENT

The principle of informed consent refers to the right of each individual potentially affected by a project to participate to an appropriate degree in decision making concerning that project. Returning to the Challenger explosion example, the astronauts should have been informed of the possibility of O-ring failure before the Challenger launch occurred.

CONFLICT OF INTEREST

Conflict of interest refers to the potential to distort good judgment while serving more than one employer or client. When this potential exists, an engineer should openly admit to these relationships in order to prevent impropriety.

ACCOUNTABILITY TO CLIENTS & CUSTOMERS

Although an engineer's primary responsibility is to protect public safety, the engineer should also perform tasks for the client or company responsibly.

FAIR TREATMENT

Engineers are entitled to a fair work environment. Employees are entitled to an environment where treatment is based on merit (nondiscrimination) and ethnic, sexual, and age harassment are not tolerated. Company policies should be spelled out in an employee handbook.

DETERMINING YOUR PERSONAL ENGINEERING ETHICS THRESHOLD FOR ACTION

During their careers, many engineers will become involved in unethical situations they cannot control. Though they choose to act responsibly—attuned to public safety, technically competent, and quickly informing their managers of positive and negative results—their managers may choose to act based on other concerns. For example, the day before the Challenger space shuttle exploded, NASA and Morton Thiokol managers decided that the O-ring data just presented were inconclusive. The launch had already been postponed by bad weather several times; launch delays had received considerable media attention because the first "teacher in space" was a member of the shuttle crew. President Reagan's State of the Union address was also scheduled for the following day. His prewritten speech contained references to the Challenger already being launched. These managers decided that the launch would proceed as scheduled.

In preparation for being involved in unethical situations you cannot control, it is important to know your limits (Figure 1.1). If one or more of the ethical dilemmas discussed in the last section occur at your place of

Figure 1.1 Dworshak Dam in Idaho. Determining your personal engineering ethics threshold for action may resemble waiting for a dam to burst.
Courtesy U.S. Army Corps of Engineers.

employment, do you still want to work there? If you personally have kept up your professional responsibilities, should you stay? Other engineering ethics textbooks, written by engineers or philosophers without industrial engineering experience, advocate internal or external whistleblowing. This is impractical advice for the rank-and-file engineer, who may be supporting a family and may be financially tied to his or her work position. Certainly, this engineer has the right to practice his profession.

WHAT IS YOUR PERSONAL THRESHOLD?

What is your personal engineering ethics threshold for action? As illustrated by the 10 anonymous case studies in Chapter 16, there are many answers to this question. In each case, an engineer was extremely troubled about an unethical work situation and resolved the situation in a unique way, whether by leaving the company, leaving the field, fighting for internal change, or minimizing interaction with the offending party. It is up to you to decide your own threshold. Because the probability is high that you may work in such an environment, it is recommended that you know your threshold before you start working full time.

REFERENCES

AACSB International, *Eligibility Procedures and Accreditation Standards for Business Accreditation.* Tampa, FL: AACSB, 2004. www.aacsb.edu/accreditation/business/standards 01–01–04.pdf.

Alford, R., 5 years after sludge spill, worry amid recovery. *AP Wire,* October 11, 2005.

Associated Press, Mine inspector settles case against MSHA. *AP Wire,* October 9, 2004.

Baura, G. D., When money wasn't king. *IEEE EMB Mag,* March/April 2005, 15–16.

Berman, D., New calling: at Bell Labs, hard times take toll on pure science. *WSJ,* A1, May 23, 2003.

Cassuto, L., Big trouble in the world of "Big Physics." *Salon,* September 16, 2002. www.salon.com.

Christensen, W. C. and Manuele, F. A., ed., *Safety Through Design.* Fairfield, NJ: American Society of Mechanical Engineers Press, 1999.

Dao, J., Mine safety official critical of policies faces firing. *NY Times,* A18, November 9, 2003.

Gioia, D. A., Pinto fires and personal ethics: a script analysis of missed opportunities. *J Bus Ethics,* 1992, 11, 379–389.

Hinman, L. M., *Ethics: A Pluralistic Approach to Moral Theory* (ed. 3). Belmont, CA: Wadsworth/Thomson Learning, 2003.

Kennedy, R. F., Jr., *Crimes Against Nature: How George W. Bush and His Corporate Pals are Plundering the Country and Hijacking Our Democracy.* New York: HarperCollins, 2004, 121–124.

Kumagai, J., The whistle-blower's dilemma. *IEEE Spectrum,* April 2004, 41, 53–55.

Schinzinger, R. and Martin, M. W., *Introduction to Engineering Ethics.* Boston: McGraw-Hill, 2000, 8, 108.

Simon, B., A toxic cover-up? *60 Minutes Television Broadcast,* April 4, 2004.

Stuart, R., Toll at 111 in Kansas City hotel disaster. *NY Times,* A1, July 19, 1981.

Unger, S. H., *Controlling Technology: Ethics and the Responsible Engineer* (ed. 2). New York: John Wiley, 1994, 109.

World Spaceflight News, *Challenger Accident: The Tragedy of Space Shuttle Flight 51-L and Its Aftermath.* Mt. Laurel, NJ: Progressive Management, 2000.

Young, S. and Berman, D., Lucent settlement unveiled by SEC: 10 face civil suits. *WSJ,* A3, May 18, 2004.

QUESTIONS FOR DISCUSSION

1. What elements of utilitarianism, duty ethics, rights ethics, and virtue ethics are present in the U.S. criminal justice system?

2. View *The Corporation,* a 2003 documentary produced by Achbar & Simpson that details the history and power of corporations. What federal agencies regulate the actions of corporations in the United States? Provide three recent examples of government regulation and, for each example, comment on the effectiveness of the regulation.

3. In *The Corporation,* one of the main interviewees is Ray C. Anderson, founder and chairman of the board of Interface, Inc., the largest commercial carpet manufacturer in the world. In 1994, Anderson pledged that his

company would become the world's first environmentally restorative company by 2020, with all inputs to its manufacturing process obtained from renewable sources and zero waste produced. Anderson is an industrial engineer by training, having received a BS in Industrial Engineering from the Georgia Institute of Technology. Read about this pledge and company mission on www.interfacesustainability.com. Critique Anderson's pledge from the perspective of the National Society of Professional Engineers Code of Ethics.

4. Software manager Mr. Incompetent is working with an external consultant, who provided the software requirements for a new cellular module within fictional Company X's current engineering project. This module will receive cellular signals and process them using digital signal processing techniques. An external consultant was used because none of the Company X software engineers has cellular or signal processing experience. The consultant's contract involves three tasks: (1) providing software requirements, (2) being available to answer questions as this software is coded, and (3) assessing the software's quality when it is completed.

The consultant recently admitted to Mr. Incompetent that he is concerned about software quality. Even though the code is not completed, the software engineer assigned to develop this code, Ms. I'm In Charge, calls the contractor frequently. The questions the engineer asks reveal she has no idea what she is coding. This is understandable, because she has no exposure to undergraduate, let alone graduate, signal processing. However, she will not admit to this lack of understanding, and the software milestones are behind schedule. The consultant has recommended that someone else code this software; worst case, he has recommendations for other software consultants. However, Mr. Incompetent considers Ms. I'm In Charge his lead engineer and believes she is the most qualified in his group. How should Mr. Incompetent bring this software task back on schedule?

5. At fictional Company Y, the Executive Committee is made up of the CEO, the chief technology officer (CTO), the chief financial officer, the vice president of operations, and the vice president of sales. The CTO is in his twilight years. He runs the development projects while his direct report, the vice president of research, Dr. Research, runs the research projects. To help keep the development projects on schedule, the CEO, who is the CTO's supervisor, decides to hire help. A new vice president of development, Mr. Ambitious, is hired, who also reports to the CTO.

One month after starting, Mr. Ambitious suggests to Dr. Research that together they could force their boss into retirement. Mr. Ambitious suggests that they go to the CEO, emphasize how the CTO is no longer needed, and ask that the CTO be forced into retirement and Dr. Research be promoted to take his place. All Mr. Ambitious asks in return is that he also be allowed to sit in on Executive Committee meetings and that a few members of Dr. Research's staff get reorganized to report to Mr. Ambitious. Dr. Research is stunned by this suggestion. What ethical dilemmas are taking place? Provide a course of action for Dr. Research.

Chapter 2

Options for Action When an Engineering Ethics Threshold Is Reached

In Chapter 1, we discussed our framework for ethical engineering decision making, which is based on ethical theories we have learned, ethical behavior we have observed, and engineering ethics codes. We advocate that engineers use this framework, as well as three professional responsibilities, to guide their professional behavior. These three responsibilities—concern for public safety, technical competence, and timely communication of positive and negative results to management—are grounded in respecting others and keeping them safe.

We also discussed ethical engineering dilemmas that we may encounter at work and asked each engineer to determine his or her personal engineering ethics threshold for action. This threshold refers to the ethical dilemmas that may occur at a workplace before an engineer is forced to an extreme action.

In Chapter 2, we discuss how to perform these extreme actions of leaving the company or fighting for change.

DEPARTURE

Once an engineer reaches his or her ethics threshold, an obvious solution is to leave. The ease with which a new job is found is affected by the economy. If at all possible, the engineer should time the job search so his or her total tenure at the current job will have been at least 12 months. Twelve months is the minimum length of time required for job duration so that managers do not consider an applicant "unstable." During the interview process, the recommendations of current colleagues are not requested. However, care should be taken in announcing departure to a new job in order to preserve recommendations for future jobs.

In announcing his or her resignation, the engineer should give a personal reason for leaving that in no way reflects on the company and should give this reason to *everyone*. One example of a successful reason I have observed is, "My wife will be giving birth soon, so I wanted to change jobs so I could spend more time with my family." Was this the true reason? More likely, the decision to leave had more to do with losing Medicare reimbursement for the company's medical products (the government reimburses physicians for medical procedures using certain types of equipment), which would eventually lead to decreasing, rather than increasing, rates of annual sales.

Standing by the stated personal reason and deflecting blame from the company improves the engineer's chance of receiving a positive recommendation later from his direct supervisor. It also prevents a decrease in employee morale. In fact, when one chief executive officer (CEO) received a candid, logical set of reasons based on company policies from one of his direct reports as to why he was leaving, the CEO asked for excuses he could provide in an e-mail explaining the direct report's departure. All of a sudden, this very urban individual was looking forward to fishing, hiking, and stargazing in his new job.

WHISTLEBLOWING

In some cases, an engineer chooses to fight an unethical situation in an attempt to correct the problem. Historically, this action has been referred to as whistleblowing. As defined by Schinzinger and Martin, "whistleblowing occurs when an employee or former employee conveys information about a significant moral problem to someone in a position to take action on the problem, and does so outside regular in-house channels for addressing disputes or grievances" (Schinzinger and Martin, 2000). When the information is conveyed to someone within the organization, it is called internal whistleblowing. When it is conveyed to someone outside the organization, it is called external whistleblowing.

Whistleblowing is an unfortunate term. As Unger observed, "It conveys the wrong impression, of someone running around, being noisy and disruptive, behaving in an erratic way. Which is the very opposite of all the engineer whistleblowers I'm aware of. They did everything they could to avoid publicity, to avoid making waves. Engineers are very quiet people" (Kumagai, 2004).

I believe a more appropriate term for whistleblower is *conscience*. Therefore, in this text we define the employee conscience as an employee working to change an organization in which he or she is employed. The employee may contact authorities within or outside the organization. We also define the observer conscience as a person working to change an organization in which he or she is not employed.

Regardless of whether an engineer decides to act within or outside an organization, there are practical procedures that should be followed. First, this action should only be performed if all normal channels have already been exhausted. During the time these normal channels are being pursued and during subsequent action, detailed records, including copies of supporting documents, should be kept of all relevant data, formal meetings, and applicable interactions. The records should stick to facts and exclude emotional observations. If possible, these actions should be conducted with other employees, as there is strength in numbers. Even if others are unwilling to join the employee, they should at least be consulted for advice so that the employee does not work in isolation. Especially if this is an external case, a lawyer should be consulted about potential legal liabilities (Schinzinger and Martin, 2000).

Realize that the reward for coming forward may be an investigation into the employee's personal and professional life. If real issues are not found, other issues may be manufactured. When Ralph Nader, in his book *Unsafe at Any Speed*, called attention to the structural defects in General Motors's Corvair, which he believed (an investigation by the National Highway Traffic and Safety Administration proved otherwise) caused the car to become uncontrollable and overturn at high speeds (see Chapter 3), General Motors hired detectives to investigate him in hopes of discrediting him. It later issued Nader a public apology and paid $425,000 to settle a civil action for invasion of privacy (Cullen, 1994).

THE EMPLOYEE CONSCIENCE

One of the most famous employee consciences in recent years is Sherron Watkins, former Enron vice president of corporate development (Figure 2.1). Her two letters to Enron Chairman Kenneth Lay detailed how Enron hid billions of dollars in debts and operating losses inside private partnerships and complex accounting schemes in order to support Enron's inflated stock price. Though Watkins herself never publicized the letters, the letters became important documentation to government investigators of the Enron scandal (Duffy, 2002). Watkins' situation and those of other employee consciences

Figure 2.1 Enron employee conscience Sherron Watkins.
Drawing by Lori Hiris from her 2003 animated film, *The Invisible Hand*. Reprinted with
permission of Lori Hiris.

laid the foundation for the Sarbanes-Oxley (SOX) Act's whistleblower
clause. In this section, we discuss applicable legislation, procedures, and
examples of employees fighting for change within an organization in which
they are employed.

EMPLOYEE PROTECTION LEGISLATION

In 1978 the U.S. Congress included language in the Civil Service Reform
Act to protect federal employees from retaliation for making disclosures of
information regarding misconduct. After the courts and government agencies
created loopholes that limited who was protected, Congress unanimously
passed the Whistleblower Protection Act (WPA) in 1989. Because the courts
and agencies continued to create exceptions for who was protected, Congress
passed amendments to strengthen the WPA in 1994. After a series of hostile
judicial rulings creating more loopholes, S.995 and H.R.3806 were introduced
to further the WPA in 2001 (POGO, 2001). Although not passed, similar

legislation was included within the Homeland Security Act of 2002. The approved provision provides federal employee and federal contractor whistleblowers the right to a legal remedy if they suffer retaliation such as loss of job or demotion (POGO, 2002).

In 2002, in response to the Enron and WorldCom scandals, Congress passed the Sarbanes-Oxley Act. This Act tightened oversight of the accounting industry, reformed securities laws, and added tough new penalties on corporate fraud. The whistleblower clause in the Act is an outgrowth of revelations that employees at Enron and WorldCom sought to warn senior management of problems with company accounting practices but were ignored. For the first time, all employees in publicly traded corporations possess comprehensive whistleblower rights:

- Comprehensive coverage for all employees of publicly traded corporations;
- Comprehensive protection for any form of harassment or discrimination;
- Broadly defined protected speech, protecting disclosures of any corporate misconduct that could threaten the value of shareholders' investments;
- Provision for administrative investigation, temporary relief, and due process hearings on alleged harassment;
- The right to a jury trial in U.S. District Court if an administrative ruling is not received in 180 days;
- For both administrative hearings and judicial trials, replacement of antiquated legal burdens of proof, in which employees only prevail on the merits in 2 to 5% of cases, with the modern standards for government workers in the WPA, in which 25 to 33% have prevailed in decisions on the merits;
- Compensatory damages and attorney fees;
- Criminal felony penalties up to 10 years for retaliation;
- Requirement for audit committees to develop complaint procedures (Devine, 2002).

The Sarbanes-Oxley Act whistleblower clause is reproduced in the Appendix.

Other federal acts, such as the Clean Water Act and the Solid Waste Disposal Act, contain whistleblower provisions. Although protection varies by state, all but 15 state governments have passed similar whistleblower protection legislation. State protection began in 1980s as a result of the erosion of the at-will employment doctrine. *At-will employment* refers to the ability of an employer to terminate employment at any time and for any reason.

EMPLOYEE PROTECTION PROCEDURES

Employee protection procedures range from corporate due process to Occupational Safety and Health Administration (OSHA) complaint filing and anonymous complaint filing.

Sarbanes-Oxley Anonymous Reporting System

An employee at a public company wishing to report dangerous behavior or illegal activity may use the company's anonymous reporting system. The Audit Committee of each publicly traded corporation was required by SOX to create a confidential, anonymous reporting system by April 2004 or be delisted. Each system includes procedures for the receipt, retention, and treatment of complaints regarding accounting, internal accounting controls, or auditing matters. Each is capable of receiving anonymous complaints from both company personnel and third parties such as competitors, vendors, and consumers.

Rather than being an ominous mandate, this reporting system is a corporate opportunity for publicly traded companies to use employees as an internal early warning system for illegal conduct and other wrongdoing. An effective program allows a company to identify illegal conduct before it occurs or before it becomes catastrophic, to correct the conduct internally, to increase accountability, and to build confidence in the company among shareholders, employees, and consumers. An effective program also precludes the wrongdoing from becoming a major focus of government enforcement agencies, Congress, or the media (Watchman, 2002).

OSHA Complaint Filing

An employee experiencing discrimination after reporting illegal activity within the organization may be eligible to file for federal action. Federal action is initiated through a complaint to OSHA, which is part of the Department of Labor. To file a complaint, the employer must have discriminated against the employee because he or she is involved in legally protected safety and health activities or because the employee reported any of the following:

- Environmental concerns;
- Potential securities fraud;
- Violations of Department of Transportation rules and regulations pertaining to commercial motor carriers;
- Violations of Federal Aviation Administration rules and regulations;
- Violations of Nuclear Regulatory Commission rules and regulations.

Discrimination may include the following actions:

- Assigning to undesirable shifts,
- Blacklisting,
- Damaging financial credit,
- Demoting,
- Denying overtime or promotion,
- Disallowing benefits,
- Disciplining,
- Evicting from company housing,
- Failing to hire or rehire,
- Firing or laying off,
- Intimidating,
- Transferring,
- Reassigning work,
- Reducing pay or hours.

The complaint should be filed as soon as possible with the local OSHA office, because legal time limits vary with each type of violation reported. If evidence supports the discrimination claim, OSHA will request that the employee's job, earnings, and benefits be restored. More information can be found at www.osha.gov (OSHA, 2003).

Anonymous Complaint Filing

In extreme cases in which reporting dangerous behavior or illegal activity within the organization is ill advised, an employee may wish to file an anonymous complaint with an external agency. If this agency can effect change in the organization, this may be the only way to halt the dangerous behavior.

EMPLOYEE PROTECTION EXAMPLES

A recent example of an anonymous corporate complaint occurred in 2000, before the anonymous SOX reporting system was mandated. Seven employees from a division of Guidant, who were later dubbed the "Anonymous Seven" in court documents (see Chapter 13), sent an anonymous letter to Guidant's chief compliance officer and the Food and Drug Administration (FDA). In this letter, they charged that their organization had failed to report numerous problems to the FDA. The letter launched an internal investigation, as well as a 3-year investigation by FDA's Office of Criminal Investigations and the FBI. In fact, more than 2500 reports of medical device complications involving the Ancure Endograft system had not been reported, though mandated by law. These hidden complications included 12 deaths. Guidant subsequently paid $92.4 million in criminal and civil penalties, the largest fine levied to date for violating the FDA's medical device reporting requirements (Jacobs, 2003).

A recent example of exposing dangerous behavior to superiors and the public occurred on October 24, 2004. On this day, Army Corp of Engineers Contracting Director Bunnatine Greenhouse sent a letter to the acting Army Secretary and copies to Congress and the news media. Greenhouse has a Bachelor of Science (B.S.) degree in mathematics and a Master of Science (M.S.) in engineering management. Her letter detailed that the Corp had shown a pattern of favoritism toward Halliburton that imperiled "the integrity of the federal contracting program."

In March 2003, Greenhouse saw no reason why the Corps awarded Halliburton subsidiary Kellogg Brown & Root (KBR), without competition, a 5-year, $7 billion contract to repair oil fields. In December 2003, Corps leaders went behind her back to issue a legal document approving the unusually high prices KBR had charged for fuel imports to Iraq. These prices are now calculated by Pentagon auditors as being inflated by at least $61 million and are the subject of criminal inquiries. In early 2004 she questioned why an expiring Halliburton logistics contract in the Balkans had to be extended from the original term of 4 years for an extra 11 months and $165 million on the grounds that no other company could do the job on time.

The Pentagon began an investigation and promised to protect Greenhouse's position (Eckholm, 2004). Halliburton has denied any wrongdoing. Exposure of alleged impropriety is not covered by OSHA whistleblower protection because OSHA complaint procedures were not followed.

THE OBSERVER CONSCIENCE

A person may also act as an observer of ethical dilemmas in an organization in which he or she is not employed. In this section, we discuss applicable legislation, procedures, and examples of persons fighting for external change.

OBSERVER PROTECTION LEGISLATION

If a citizen has evidence of fraud (excluding tax fraud) against government contracts and programs, the citizen may sue, on behalf of the government, in order to recover the stolen funds. The violator is liable for three times the dollar amount the government was defrauded and for civil penalties of

$5500 to $11,000 for each false claim. In compensation for the risk and effort of filing a qui tam case, the observer conscience is rewarded a portion of the funds recovered, typically 15 to 25%. The term *qui tam* comes from the Latin phrase, "*Qui tam pro domino rege quam pro sic ipso in hoc parte sequitur,*" which translates to, "He who sues for the king in this matter sues for himself as well."

The qui tam section is the first section of the False Claims Act, a Civil War–era law that Congress rejuvenated in 1986 with amendments in order to fight rampant government fraud. In the second section of this Act, discharge or harassment of the citizen filing the qui tam suit is prohibited. In response to discharge or harassment, the citizen may file a wrongful discharge suit for double back pay and other damages. This antiretaliation provision was modeled after other whistleblower laws.

From 1986 to 2003, False Claims Act settlements and judgments totaled more than $12 billion. In fiscal year 2003 alone, a record $2.1 billion was recovered under the False Claims Act (FCALC, 2004).

OBSERVER PROTECTION PROCEDURES

Observer protection procedures range from False Claims Act procedures to anonymous complaint filing.

False Claims Act Complaint Filing

A qui tam complaint must be filed within the later of two time periods: (1) 6 years from the date of the violation of the Act, or (2) 3 years after the government knows or should have known about the violation, but in no event longer than 10 years after the violation of the Act. The citizen filing must be the first to file for this violation. The complaint must be in federal district court, in accordance with the Federal Rules of Civil Procedure. Additionally, a copy of the complaint, along with a written disclosure statement of all material evidence the citizen possesses, must be served on the attorney general of the United States and should be served on the U.S. attorney for the district in which the action is brought.

The complaint will remain in strict confidence for at least 60 days, during which time the government will investigate allegations. If the government intervenes and proceeds, the Department of Justice will have the primary responsibility for prosecuting the case. The time from government intervention to case settlement varies; some cases are settled within 1 year (FCALC, 2004).

Anonymous Complaint Filing

Alternatively, an observer may wish to file an anonymous complaint with an external agency to report dangerous or illegal behavior in an organization. If this agency can effect change in the organization, this may be the only way to halt the dangerous behavior.

OBSERVER PROTECTION EXAMPLES

An anonymous shipment of 4000 pages of tobacco company documents became key evidence in the state of Mississippi's lawsuit against the tobacco industry to recover $940 million the state had spent treating sick smokers. The shipment was sent by Merrell Williams, a former paralegal who spent 4 years illegally copying Brown and Williamson documents. At the time, Williams was employed at the Louisville, Kentucky, law firm of Wyatt Tarrant & Combs, which represented Brown and Williamson. His assignment during his employment from 1988 to 1992 was to identify internal Brown and Williamson records and research findings that might be helpful to people suing the tobacco industry, should the records come out in court. Williams later stated his purpose in stealing the documents "was to change a perspective, an idea in the world, which has a long tradition" (Nguyen, 2004).

Among other disclosures, the documents proved that Brown and Williamson had known as early as 1963 that nicotine was addictive. This contrasted starkly with 1994 testimony by tobacco CEOs and presidents from seven companies that nicotine was not addictive (Frontline, 1998). On July 4, 1997, four tobacco companies agreed to pay Mississippi $3.4 billion over 25 years but did not admit liability. This was the first formal settlement of lawsuits between tobacco companies and any of 40 states that filed actions against them (Meier, 1997).

A recent example of a successful qui tam complaint was filed by a Catholic priest and his colleague in July 2002. Rev. John Corapi was seen by Dr. Chae Moon in Redding, California, who performed cardiac catheterization and recommended that Corapi undergo immediate multiple-vessel heart bypass surgery at Redding Medical Center (RMC). Because Corapi believed the surgery was unnecessary, he obtained second opinions on the procedure outside the Redding medical system. After complaining unsuccessfully to officials at RMC about the inaccurate surgical recommendation, Corapi, along with colleague Zerga, reported the complaint to the FBI, which launched an investigation.

FBI agents raided RMC on October 30, 2002, and removed applicable medical and billing records. Less than 1 week after the raid, Corapi and

Zerga filed a sealed civil complaint in Sacramento federal court alleging fraud on the part of Tenet Heathcare Corporation, which owned RMC; Dr. Moon; Dr. Fidel Realyvasquez, a cardiac surgeon at the hospital; and Cardiology Associates of Northern California, a group that included Moon. FBI investigation, which included interviews with hospital staff, revealed that a substantial number of unnecessary surgeries had been charged to Medicare; $54 million was recovered in the settlement with Tenet only, which admitted no wrongdoing. Corapi and Zerga received 15% of the total recovered, which amounted to $8.1 million. Tenet still faces more than 100 medical malpractice suits on behalf of RMC patients and an ongoing investigation of its practices by the U.S. Senate Finance Committee (Walsh, 2004).

CONCLUSION

Not all corporations are unethical. Some are even shining examples of social responsibility. Part of Google's corporate philosophy is, "You can make money without doing evil" (Google, 2004). One percent of Google profits go to the charitable Google Foundation. Birkenstock USA has been doing good works for 30 years but is reluctant to disclose the recipients of its philanthropy (Lewis, 2004). Perhaps readers may work at these types of corporations. With SOX protections in place and more False Claims complaints being filed, I believe that workplaces are becoming more ethical. But if ethical dilemmas occur, then at least the alternatives for action have been discussed.

REFERENCES

Cullen, F., Maakestad, W., and Cavender, G., Profits vs. Safety. Reprinted in *The Ford Pinto Case: A Study in Applied Ethics, Business, and Technology*. Edited by D. Birsch and J. H. Fielder. Albany, NY: SUNY Press, 1994, 263–272.

Devine, T., *Sarbanes-Oxley Act Summary of Law*. Washington, DC: Government Accountability Program, 2002. http://www.whistleblower.org/corporate/sox_summary.htm.

Duffy, M., By the sign of the crooked E. *Time Mag*, Web exclusive, January 19, 2002. http://www.time.com/time/business/article/0,8599,195268,00.html.

Eckholm, E., A watchdog follows the money in Iraq. *NY Times*, A10, November 15, 2004.

False Claims Act Legal Center, *Taxpayers Against Fraud* Web site. Washington, DC: False Claims Act Legal Center, 2004. www.taf.org.

Frontline, *Inside the Tobacco Deal: How Two Small-Time Lawyers from Mississippi Took Big Tobacco to the Edge of Bankruptcy and Criminal Prosecution*. Washington, DC: Public Broadcasting System, 1998. http://www.pbs.org/wgbh/pages/frontline/shows/settlement.htm.

Google, *Our Philosophy*, Mountain View, CA: Google, 2004. http://www.google.com/corporate/tenthings.html.

Herper, M., Face of the Year: David Graham. *Forbes*, December 13, 2004. http://www.forbes.com/home/sciencesandmedicine/2004/12/13/cx_mh_1213faceoftheyear.html.

Jacobs, P., Medical firm's dangerous secret device's troubles were well known at Menlo Park company. *San Jose Mercury News*, 1A, August 3, 2003.

Kumagai, J., The whistle-blower's dilemma, *IEEE Spectrum*, April 2004, 41, 53–55.

Lewis, M., The irresponsible investor. *NY Times Mag*, 68, June 6, 2004.

Meier, B., Acting alone, Mississippi settles suit with 4 tobacco companies. *NY Times*, A1, July 4, 1997.

Nguyen, T., Refiled under "U": Taking a page from history? Try not to sweat. *Wash Post*, D01, July 25, 2004.

NOW, David Graham Transcript, *NOW on PBS Television Program*, January 7, 2005. http://www.pbs.org/now/transcript/transcriptNOW101_full.html.

Occupational Safety and Hazard Administration, *Whistleblower Protection OSHA Fact Sheet*. Washington, DC: OSHA, Department of Labor, 2003.

Project on Government Oversight (POGO), *Media Briefer: Whistleblower Protection Act*. July 7, 2001. http://www.pogo.org/p/government/ga-010707-whistleblower.html.

Project on Government Oversight, *Whistleblower Protections Included On Homeland Security Bill*. July 12, 2002. http://www.pogo.org/p/government/ ga-020724-whistleblower.html.

Schinzinger, R. and Martin, M. W., *Introduction to Engineering Ethics*. Boston: McGraw-Hill, 2000, 167–168, 178.

Walsh, D., Whistle-blowers get $8.1 million, *Sacramento Bee*, B1, January 8, 2004.

Watchman, G. R., *Sarbanes-Oxley Whistleblowers: A New Corporate Early Warning System*. Washington, DC: Government Accountability Program, 2002. http://www.whistleblower.org/corporate/sox_whistleblowers_analysis_gwatchman.pdf

QUESTIONS FOR DISCUSSION

1. In December 2002, *Time* magazine named Sherron Watkins and two other women, Cynthia Cooper and Coleen Rowley, as their "Persons of the Year." What did Ms. Cooper uncover? How was she treated by fellow employees? How did her revelation affect WorldCom?

2. View *The Insider*, a 1999 feature film produced by Brugge that details how investigative journalism was involved in exposing corporate tobacco dishonesty. Insider conscience Dr. Jeffrey Wigand was portrayed by Russell Crowe. What were Wigand's motivations for disclosing confidential Brown and Williamson information? Did these motivations decrease the credibility of his information?

3. In the summer of 2004, in the midst of emerging evidence that the drug Vioxx increased the risk of heart attack or stroke, Dr. David Graham was preparing to give his own talk on Vioxx at an international pharmacoepidemiology conference in France. Graham, an associate director of science at the FDA Office of Drug Safety, had conducted a study that demonstrated that both low and high doses of Vioxx increased the risk of heart attack compared with another drug in the same class, Celebrex.

His managers began to pressure him to relax his conclusions. When one manager insisted that industry was his client, he replied that "industry may be your client, but it will never be my client" (NOW, 2005).

After Merck voluntarily withdrew Vioxx on September 30, 2004, Senate Finance Committee Chair Chuck Grassley began an investigation into Vioxx. Graham was asked to testify. For testifying that Vioxx was not an isolated accident, but part of a systemic failure at FDA to address drug safety that Graham had personally witnessed over his 20-year career at the agency, *Forbes* magazine named Graham their Face of the Year (Herper, 2004). Name the ethical dilemmas discussed in Chapter 1 that Graham experienced.

4. How effective do you believe the SOX anonymous reporting system is? What evidence do you have to support your belief?
5. Find the top 20 False Claims Act judgments and settlements. What industry dominates these fraud judgments? Why do you believe this occurred?

Part II

National Case Studies

September 2002 to August 2003 was a terrible 12-month period in engineering ethics. During this time, a scientist at Bell Laboratories committed nanotechnology fraud, the Columbia space shuttle exploded, Guidant received the largest pay-out to date from the Food and Drug Administration for hiding medical device defects, and the New York City blackout occurred. While reflecting on these events, I wondered how often disasters occurred after warnings by engineers. I knew that warnings about the Columbia explosion mirrored the warnings about the Challenger explosion. As an experiment, I challenged myself to find similar major disasters. Within 4 hours of Googling, I found eight other national examples. Later, in early 2005, I replaced the eighth example with the Indian Ocean tsunami of 2004.

Each of these 13 disasters is detailed in a case study chapter. Each case study chapter contains the following sections: the news story as reported by the *New York Times*, the back story, applicable regulations, the engineering warning, and discussion questions. Case analysis is used in programs such as Harvard Business School because it prepares students to deal with the types of problems encountered in professional practice. Using an ethical vocabulary based on previous experience and study, we discuss the decisions made by engineers and their managers that eventually made headlines. The discussion questions at the end of each chapter are intentionally vague in order to stimulate discourse. By practicing ethical analysis, we strengthen our ability to conduct it.

National Case Studies

Chapter 3

1978: Ford Pinto Recall

THE REPORTED STORY

The *New York Times* Abstract:

After months of vigorously defending the safety of the fuel systems in its Pinto automobile, the Ford Motor Company today announced the recall of 1.5 million of the subcompacts for "modifications" of the fuel system aimed at increasing resistance to leakage and diminishing the risk of fire in the event of rear-end crashes. (Stuart, 1978)

THE BACK STORY

AUTOMOBILE SAFETY

When Henry Ford began to market the Model T, the first mass-produced automobile, in 1908, its design had no specific provisions for safety. Until 1955, safety door locks were not installed in any model, even though doors opened in 42% of all serious crashes until that time. In general, the public believed that the primary cause of accidents was improper driving. The entire safety establishment, which was heavily influenced by the auto industry, promoted the view that safety meant "safe behavior by drivers" (Dirksen, 1997).

The Ford Motor Company attempted to change this viewpoint in 1956, when it introduced its "Lifeguard Design." This new design involved equipping cars with a deep-dish steering wheel, padded seatbacks, swing-away rearview mirrors, safety door latches, safety belts, padded dashboards, and padded sun visors and rearview mirrors (Dirksen, 1997). The intent of these features was to minimize the effect of a driver colliding with the inside of his or her car during a crash. Unfortunately, General Motors (G.M.) Chevrolet's emphasis on new design and a more powerful V-8 engine led to more sales than Ford that year, leading many in the auto industry to conclude that "safety doesn't sell." Ford management responded by deemphasizing safety and focusing more on style and horsepower (Fielder, 1994).

Ralph Nader challenged this viewpoint in 1965 with the publication of his book *Unsafe At Any Speed*. He called attention to the structural defects in G.M.'s Corvair, which he believed (an investigation by the National Highway Traffic and Safety Administration proved otherwise) caused the car to become uncontrollable and overturn at high speeds. He also raised the question, Who is responsible for injuries and deaths due to auto accidents? Said Nader, "The prevailing view of traffic safety is much more a political strategy to defend special interests than it is an empirical program to save lives and prevent injuries . . . Under existing business values potential safety advances are subordinated to other investments, priorities, preferences, and themes designed to maximize profit" (Nader, 1965).

In response, G.M. hired detectives to investigate Nader in hopes of discrediting him. It later issued Nader a public apology and paid $425,000 to settle a civil action for invasion of privacy. These disclosures caused great public outrage and put pressure on the U.S. Congress to pass the Highway Safety Act and the National Traffic and Motor Vehicle Safety Act in 1966. During this time, the annual death rate from auto accidents approached 50,000. The Highway Safety Act mandated federal regulation of the automotive industry and led to the creation of an enforcement agency, the National Highway Traffic Safety Administration (NHTSA) (Cullen, 1994). The Motor Vehicle Safety Act required NHTSA to issue new safety regulations by January 31, 1967.

Ford management met with NHTSA in 1966 to convince the agency that auto accidents are caused by "people and highway conditions" (Dowie, 1994). One result of this meeting was an informal agreement that cost-benefit analysis could be used by auto manufacturers to determine auto safety decisions. Cost-benefit analysis is a business tool used to determine whether the cost of a project justifies the dollar value of benefits that would be derived. Cost-benefit analysis was first used at Ford by Robert McNamara, who eventually became Ford president. After McNamara left Ford to become Secretary of Defense under Presidents Kennedy and

Johnson, he implemented many government applications of cost-benefit analysis (Dowie, 1994).

Standard 301, which deals with fuel spill standards in accidents, was first proposed by NHTSA in 1968 (Fielder, 1994). When it was first proposed, Standard 301 had the potential to delay market release of Ford's new subcompact car.

FORD SUBCOMPACT CAR PROJECT

Ford's new subcompact car project was spearheaded by Lee Iacocca. Originally an engineer with a Master's degree in Engineering from Princeton, Lee Iacocca quickly shifted from engineering to sales when he was hired by Ford after graduation in 1946. By the mid 1960s, he was known as the father of the Mustang, having led the project that resulted in market release of the Mustang in 1964. Iacocca forcefully argued to Chief Executive Officer (CEO) Henry Ford II that the Germans and Japanese would capture the entire American subcompact market unless Ford released its own alternative to the Volkswagen Beetle. Because Iacocca wanted this car in American showrooms with the 1971 models, he ordered his car engineering vice president, Robert Alexander, to oversee the shortest production planning period at that time. Rather than spending the typical 43 months from conception to production, the Pinto schedule was set to just under 25 months (Figure 3.1). Iacocca also set an important goal he called "the limits of 2000." The Pinto weight limit was 2000 pounds; the Pinto cost limit was $2000 (Dowie, 1994).

Typically, marketing surveys and preliminary engineering precede the styling of a new auto model. However, with such a short schedule, styling preceded most of the engineering and dictated engineering design (see Figure 3.1). Because of styling constraints, locating the gas tank over the axle, which was known to prevent fire in rear-end crashes, was undesirable. The axle arrangement, in concert with styling constraints, resulted in a small luggage compartment that would be limited in carrying long objects such as golf clubs. To increase the size of the luggage compartment, the gas tank was relocated to the car's rear (Strobel, 1994) (Figure 3.2).

Because tooling (the production of equipment used in manufacturing processes) had a fixed time frame of 18 months, it began shortly after design. With $200 million invested in tooling, even poor crash test results, which should have triggered major gas tank redesign, did not delay the Pinto project schedule. The Pinto was not able to pass part of proposed (but not yet implemented) Standard 301, which limited fuel spillage to 1 ounce per minute when rear-ended by a barrier moving at 30 miles

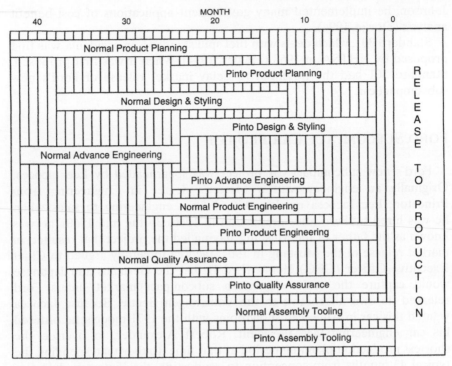

Figure 3.1 Automobile preproduction schedule.
Based on Dowie, 1994. Courtesy *Mother Jones* magazine.

per hour. Billed as "the carefree little American car," the Pinto retailed for $1919 when it was released on September 11, 1970. The price was about $170 less than the price announced for its soon-to-be released competitor, the Chevrolet Vega, and within $80 of the bestselling Volkswagen Beetle (Dowie, 1994).

As shown in Figure 3.2, the Pinto fuel tank was constructed of sheet metal and was attached to the auto undercarriage by means of two metal straps. The fuel filler pipe, which transported pumped gas to the fuel tank, was affixed to the inner side of the left rear quarter panel by means of a bracket firmly attached to the quarter panel surface. The fuel filler pipe extended into the top left side of the tank in a sliding fit through a sealed opening. The fuel tank held approximately 11 gallons for engine operation. With sufficient rear impact, the fuel filler pipe was completely dislodged from the tank, causing fuel spillage in a wide dispersive fashion. In impacts sufficient to puncture or tear the fuel tank, fuel spillage occurred in a pouring fashion (NHTSA, 1994a).

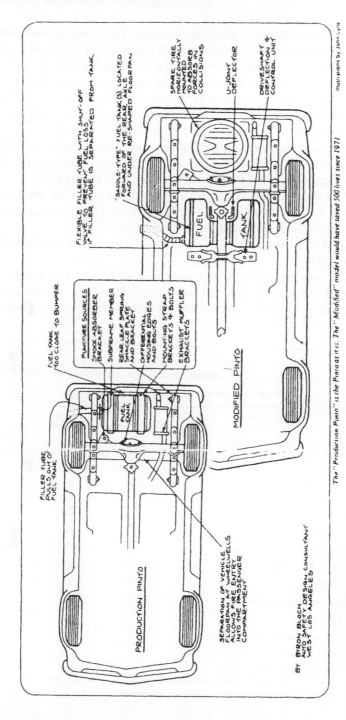

Illustrations by John Lytle

FLEXIBLE FILLER TUBE WITH SHUT-OFF VALVE TO PREVENT FUEL LOSS IF FILLER TUBE IS SEPARATED FROM TANK

SPARE TIRE HORIZONTALLY MOUNTED TO ABSORB FORCES IN COLLISIONS

U-JOINT DEFLECTOR

DRIVESHAFT DEFLECTION & CONTROL UNIT

"SADDLE-TYPE" FUEL TANK(S) LOCATED FORWARD OF THE REAR AXLE AND UNDER RE-SHAPED FLOORPAN

FUEL TANK

MODIFIED PINTO

The "Production Pinto" is the Pinto as it is. The "Modified" model would have saved 500 lives since 1971

FILLER TUBE PULLS OUT OF FUEL TANK

PUNCTURE SOURCES

SHOCK ABSORBER BRACKET

SUBFRAME MEMBER

REAR LEAF SPRING SHACKLE PLATE AND BRACKET

DIFFERENTIAL HOUSING EDGES AND BOLTS

MOUNTING STRAP BRACKETS & BOLTS

EXHAUST MUFFLER BRACKETS

FUEL TANK TOO CLOSE TO BUMPER

FUEL TANK

PRODUCTION PINTO

SEPARATION OF VEHICLE FLOORPAN AT WHEELWELLS ALLOWS FIRE ENTRY INTO THE PASSENGER COMPARTMENT

BY BYRON BLOCH
AUTO SAFETY DESIGN CONSULTANT
WEST LOS ANGELES

Figure 3.2 The "production" Pinto and a safer modified model. From Dowie, 1994. Courtesy *Mother Jones* magazine.

In order to justify the rear gas tank location, Ford first argued, when Standard 301 was proposed in 1968, that fire was a minor problem in auto crashes. This caused NHTSA to contract several independent research groups to study auto fires. Robert Nathan and Associates found that 400,000 autos were burning each year, burning to death more than 3000 people. Ford lobbyists then argued that while burn accidents did occur, rear-end collisions were relatively rare. After another round of analysis, NHTSA determined that rear-end collisions were seven and a half times more likely to result in fuel spills than were front-end collisions. By 1972 these delay tactics had stalled passage of Standard 301 for 4 years.

When NHTSA determined in a 1972 report that human life was worth $200,725 (Table 3.1), Ford rounded the figure off to $200,000 and conducted a cost-benefits analysis for redesigning the Pinto.

It determined that the cost of $137 million far outweighed the benefit of $49.5 million (Table 3.2).

The analysis was based on a unit auto cost of $11 to strengthen gas tank integrity. Ford further delayed passage of Standard 301 by stating that rear-end collision deaths were caused by the kinetic force of impact,

Table 3.1

Societal Cost Components for Fatalities

Component	1971 Costs
Future Productivity Losses	
Direct	$132,000
Indirect	41,300
Medical Costs	
Hospital	700
Other	425
Property Damage	1,500
Insurance Administration	4,700
Legal and Court	3,000
Employer Losses	1,000
Victim's Pain and Suffering	10,000
Funeral	900
Assets (Lost Consumption)	5,000
Miscellaneous Accident Cost	200
TOTAL PER FATALITY:	**$200,725**

Adapted from Dowie, 1994. Courtesy *Mother Jones* magazine.

Table 3.2

Ford Cost-Benefit Analysis for Fuel Leakage

	Benefits	Costs
Savings/Sales	180 burn deaths 180 serious burn injuries 2,100 burned vehicles	11 million cars 1.5 million light trucks
Unit Cost	$200,000 per death $67,000 per injury $700 per vehicle	$11 per car $11 per truck
TOTAL:	**$49.5 million**	**$137 million**

Adapted from Dowie, 1994. Courtesy *Mother Jones* magazine.

not burns. After NHTSA again commissioned studies to analyze impacts versus burns, the Insurance Institute for Highway Safety determined through careful study that corpses taken from burned autos in rear-end crashes contained no cuts, bruises, or broken bones. These corpses would have survived the accident unharmed if the auto had not ignited. Ford also complained about the test conditions described in Standard 301.

Ford's arguments contributed to the passage of Standard 301 being delayed for 8 years in total, until 1976, during which time more than two million Pintos were manufactured (Dowie, 1994). The 1977 Pinto was the first model equipped with a protected fuel tank, prompted by the adoption of Standard 301.

PINTO INVESTIGATIONS

In 1977, Mark Dowie exposed the tendency of Ford Pintos to ignite during rear-end collisions and Ford's attempt to delay passage of a standard the Pinto could not meet. In his article for *Mother Jones* magazine, Dowie accused Ford of causing 500 to 900 burn deaths because it was unwilling to pay $11 more per vehicle for a safer gas tank (Dowie, 1994). This prompted NHTSA to investigate the Pinto's safety on September 13, 1977.

The results of NHTSA's investigation were released in May 1978. Investigation results were based on reports from consumers and Ford, examination of accident statistics, crash test results from tests commissioned by NHTSA, and motor vehicle record checks. NHTSA observed that "the fuel tank and filler pipe assembly installed in the 1971–1976 Ford Pinto is subject to damage which results in fuel spillage and fire potential in rear impact collisions by other vehicles at moderate closing speeds" (NHTSA, 1994a). The investigation conclusions are listed below:

1. 1971–1976 Ford Pintos have experienced moderate speed, rear-end collisions that have resulted in fuel tank damage, fuel leakage, and fire occurrences that have resulted in fatalities and non-fatal burn injuries.
2. Rear-end collision of Pinto vehicles can result in puncture and other damage of the fuel tank and filler neck, creating substantial fuel leakage, and in the presence of external ignition sources fires can result.
3. The dynamics of fuel spillage are such that when impacted by a full size vehicle, the 1971–1976 Pinto exhibits a "fire threshold" at closing speeds between 30 and 35 miles per hour.
4. Relevant product liability litigation and previous recall campaigns further establish that fuel leakage is a significant hazard to motor vehicle safety, including such leakage which results from the crashworthiness characteristics of the vehicle.
5. The fuel tank design and structural characteristic of the 1971–1976 Mercury Bobcat which render it identical to contemporary Pinto vehicles, also render it subject to like consequences in rear impact collisions. (NHTSA, 1994a)

One month after publication of the NHTSA report, Ford recalled 1.5 million Pintos and Mercury Bobcats, which had similar fuel systems. It replaced the filler pipe and added two polyethylene shields to help protect the tank. Ford estimated the recall cost $20 million after taxes (Strobel, 1994).

Pinto Lawsuits

Numerous lawsuits were filed against Ford Motor Company by Pinto burn victims. The number filed has been estimated to range from several dozen to more than 100 lawsuits. Two lawsuits greatly affected Ford. In 1978 a jury awarded a Pinto burn victim $125 million in punitive damages. Although the damages were later reduced to $6.6 million, a judgment upheld on appeal prompted the appeals judge to assert "Ford's institutional mentality was shown to be one of callous indifference to public safety."

On August 10, 1978, three teenage girls died in a fire triggered after their 1973 Pinto was rear-ended by a van. Witnesses claimed to have seen a relatively low-speed collision. A grand jury indicted Ford on charges of reckless homicide. This was the first time a corporation was tried for criminal behavior. In order to prevent a legal precedent for all manufacturing industries, Ford assembled a defense team led by Watergate prosecutor James Neal. During the ensuing media trial, the defense convinced the jury that the Pinto involved in the accident was stopped when it was hit by the van. Therefore a low-speed collision did not occur, and the deaths were not the result of reckless homicide. Although Ford was found innocent, the reputation of the Pinto was forever harmed by the trial (Gioia, 1994).

APPLICABLE REGULATIONS

Portions of Motor Vehicle Safety Standard, Part 571; S301, are listed below. The rear-moving barrier crash requirement that the Pinto did not pass is S6.2. In this passage, GVWR stands for gross vehicle weight rating.

S1. Scope. This standard specifies requirements for the integrity of motor vehicle fuel systems.

S2. Purpose. The purpose of this standard is to reduce deaths and injuries occurring from fires that result from fuel spillage during and after motor vehicle crashes.

S3. Application. This standard applies to passenger cars, and to multipurpose passenger vehicles, trucks, and buses that have a GVWR of 10,000 pounds or less and use fuel with a boiling point above 32° F, and to school buses that have a GVWR greater than 10,000 pounds and use fuel with a boiling point above 32° F.

S4. Definition. "Fuel spillage" means the fall, flow, or run of fuel from the vehicle but does not include wetness resulting from capillary action.

S5. General requirements.

S5.1 Passenger cars. Each passenger car manufactured from September 1, 1975, to August 31, 1976, shall meet the requirements of S6.1 in a perpendicular impact only, and S6.4. Each passenger car manufactured on or after September 1, 1976, shall meet all the requirements of S6, except S6.5.

. . .

S5.5 Fuel spillage: Barrier crash. Fuel spillage in any fixed or moving barrier crash test shall not exceed 1 ounce by weight from impact until motion of the vehicle has ceased, and shall not exceed a total of 5 ounce by weight in the 5-minute period following cessation of motion. For the subsequent 25-minute period (for vehicles manufactured before September 1, 1976, other than school buses with a GVWR greater than 10,000 pounds: the subsequent 10-minute period), fuel spillage during any 1-minute interval shall not exceed 1 ounce by weight.

S5.6 Fuel spillage: Rollover. Fuel spillage in any rollover test, from the onset of rotational motion, shall not exceed a total of 5 ounces by weight for the first 5 minutes of testing at each successive 90° increment. For the remaining testing period, at each increment of 90°, fuel spillage during any 1-minute interval shall not exceed 1 ounce by weight.

S6. Test requirements. Each vehicle with a GVWR of 10,000 pounds or less shall be capable of meeting the requirements of any applicable barrier crash test followed by a static rollover, without alteration of the vehicle during the test sequence. A particular vehicle need not meet further requirements after having been subjected to a single barrier crash test and a static rollover test.

S6.1 Frontal barrier crash. When the vehicle traveling longitudinally forward at any speed up to and including 30 mph impacts a fixed collision barrier that is perpendicular to the line of travel of the vehicle, or at any angle up to 30° in either direction from the perpendicular to the line of travel of the vehicle, with 50th-percentile test dummies as specified in Part 572 of this chapter at each front outboard designated seating position and at any other position whose protection system is required to be tested by a dummy under the provisions of Standard No. 208, under the applicable conditions of S7, fuel spillage shall not exceed the limits of S5.5 (Effective: October 15, 1975)

S6.2 Rear moving barrier crash. When the vehicle is impacted from the rear by a barrier moving at 30 mph, with test dummies as specified in Part 572 of this chapter at each front board designated seating position, under the applicable conditions of S7, fuel spillage shall not exceed the limits of S5.5.

S6.3 Lateral moving barrier crash. When the vehicle is impacted laterally on either side by a barrier moving at 20 mph with 50th-percentile test dummies as specified in Part 572 of this chapter at positions required for testing to Standard No. 208, under the applicable conditions of S7, fuel spillage shall not exceed the limits of S5.5.

S6.4 Static rollover. When the vehicle is rotated on its longitudinal axis to each successive increment of 90°, following an impact crash of S6.1, S6.2, or S6.3, fuel spillage shall not exceed the limits of S5.6.

S6.5 Moving contoured barrier crash. When the moving contoured barrier assembly traveling longitudinally forward at any speed up to and including 30 mph impacts the test vehicle (school bus with a GVWR exceeding 10,000 pounds) at any point and angle, under the applicable conditions of S7.1 and S7.5, fuel spillage shall not exceed the limits of S5.5 (NHTSA, 1994b).

AN ENGINEERING PERSPECTIVE

In 1968, Ford was aware that the threat of fire in rear-end crashes could be reduced using relatively inexpensive fuel system design considerations. It had partially financed a study by UCLA researchers that had come to this conclusion. The study recommended that the fuel tank not be located directly adjacent to the bumper but moved above the rear axle.

In early 1969, 1½ years before the Pinto was introduced, Ford engineers took three Ford Capris and modified their rears to be similar to the proposed Pinto. For these tests, the fuel tank was moved from above the rear axle to the rear. When one was backed into a wall at 17.8 miles per hour, the welds on the gas tank split open, the tank was damaged when it hit the axle, the filler pipe pulled out, and the tank fell out of the car, resulting in massive gas spillage. Because the welds on the car's floor split open, gasoline could spill into the car interior. In two other tests, a car was rear-ended by moving barriers at 21 miles per hour. This caused gas to leak either from the filler pipe pulling out or from the punctured fuel tank (Strobel, 1994).

Even still, the engineers responsible for Pinto components signed off approval to their immediate supervisors. The Pinto crash tests were forwarded up the chain of command to the regular product meeting chaired by Robert Alexander, vice president of car engineering, and Harold MacDonald, group vice president of car engineering. Harold Copp, a former executive in charge of the crash testing program, testified that the highest level of Ford management decided to produce the Pinto, knowing that the Pinto could ignite during low-speed rear-end collisions and that design fixes were feasible at nominal cost (West's California Reporter, 1994).

Within a few months of the Pinto's release on September 11, 1970, a standard Pinto was crashed backward into a concrete wall at 21 miles per hour. In a report marked "confidential," engineer H. P. Snider reported that the Pinto's soft rear-end crushed 18 inches in 91 msec. According to Snider, "The filler pipe was pulled out of the fuel tank and fluid discharged through the outlet. Additional leakage occurred through a puncture in the upper right front surface of the fuel tank which was caused by contact between the fuel tank and a bolt on the differential housing" (Strobel, 1994). Additionally, the tank was punctured twice by nearby metal objects and both passenger doors jammed shut, which would have prevented quick escape or rescue during a crash. This pattern of gasoline spillage remained consistent in other crash tests at lower speeds. On December 15, 1970, when a Pinto was rear-ended by a moving barrier at 19.5 miles per hour, the filler pipe pulled out, causing gas to escape and the left door to jam shut.

Later, in early 1971, Ford engineers investigated various design changes to improve crash test results. With a heavy rubber bladder reinforced with nylon lining the metal gas tank, gasoline did not spill during a 26-mile-per-hour crash into a cement wall. The bladder was estimated to cost $6 per car. An alternative liner of polyurethane foam between the inner and outer metal fuel tank shells was estimated at $5 per car. To prevent fuel tank puncture by the differential housing, engineers suggested an ultra-high-molecular poly-ethylene shield, which was estimated to cost $0.22 per car. Normally, a Pinto would have been extensively damaged and spilled gas when crashed backward into a cement wall at 21 miles per hour. However, with the addition of two side rails, it sustained considerably less damage and did not leak gas. These side rails were estimated to cost $2.40 per car. Unfortunately, Ford executives decided against adopting any of these design changes. (Design changes require signature approval at several levels of management.) An October 26, 1971 memo labeled "confidential" documented that there would be no additional improvements for the 1973 and later models of the Pinto until "required by law" (Strobel, 1994).

REFERENCES

Chen, D. W., What's a life worth? *NY Times*, D4, June 20, 2004.

Cullen, F., Maakestad, W., and Cavender, G., Profits vs. safety. Reprinted in *The Ford Pinto Case: A Study in Applied Ethics, Business, and Technology*. Edited by D. Birsch and J. H. Fielder. Albany, NY: SUNY Press, 1994, 263–272.

Dirksen, S. *History of Automobile Safety*. Smithfield, RI: Bryant University History of American Technology Class Project, 1997. http://web.bryant.edu/~history/h364proj/sprg_97/dirksen/backgrou.html.

Dowie, M., Pinto madness. Reprinted in *The Ford Pinto Case: A Study in Applied Ethics, Business, and Technology*. Edited by D. Birsch and J. H. Fielder. Albany, NY: SUNY Press, 1994, 15–36.

Fielder, J. H., The ethics and politics of automobile regulation. Reprinted in *The Ford Pinto Case: A Study in Applied Ethics, Business, and Technology*. Edited by D. Birsch and J. H. Fielder. Albany, NY: SUNY Press, 1994, 285–301.

Gioia, D., Pinto fires and personal ethics: A script analysis of missed opportunities. Reprinted in *The Ford Pinto Case: A Study in Applied Ethics, Business, and Technology*. Edited by D. Birsch and J. H. Fielder. Albany, NY: SUNY Press, 1994, 97–116.

Nader, R., *Unsafe At Any Speed: The Designed-In Dangers of the American Automobile*. New York: Grossman, 1965, 236.

National Highway Traffic Safety Administration, Investigation report, phase I: Alleged fuel tank and filler neck damage in rear-end collision of subcompact passengers cars 1971–1976 Ford Pinto 1975–1976 Mercury Bobcat. Reprinted in *The Ford Pinto Case: A Study in Applied Ethics, Business, and Technology*. Edited by D. Birsch and J. H. Fielder. Albany, NY: SUNY Press, 1994a, 77–95.

National Highway Traffic Safety Administration, Motor vehicle safety standard, Part 571; S301. Reprinted in *The Ford Pinto Case: A Study in Applied Ethics, Business, and Technology*. Edited by D. Birsch and J. H. Fielder. Albany, NY: SUNY Press, 1994b, 61–75.

Strobel, L., The Pinto documents. Reprinted in *The Ford Pinto Case: A Study in Applied Ethics, Business, and Technology*. Edited by D. Birsch and J. H. Fielder. Albany, NY: SUNY Press, 1994, 41–53.

Stuart, R., Ford orders recall of 1.5 million Pintos for safety changes: inquiry begun last fall. *NY Times*, A1, June 10, 1978.

West's California Reporter, The Pinto fuel system. Reprinted in *The Ford Pinto Case: A Study in Applied Ethics, Business, and Technology*. Edited by D. Birsch and J. H. Fielder. Albany, NY: SUNY Press, 1994, 55–60.

QUESTIONS FOR DISCUSSION

1. Should cost-benefit analysis include the costs of legal settlements and equipment recalls? What other factors could be considered in this analysis?

2. View *The Fog of War,* a 2003 documentary about Robert McNamara produced by Morris, Williams, and Ahlberg. Robert McNamara, who received a Master of Business Administration (MBA) from Harvard, viewed the world's problems as solvable through statistical analysis. Of which ethical theory does this remind you? Based on the film, how did McNamara use cost-benefit analysis in his decisions (1) on bombing raids over Japan during World War II? (2) on safety at Ford Motor Company? (3) on support for the Vietnam War as U.S. Secretary of Defense? Do you agree with each decision?

3. According to the *New York Times,* the median payment of families of September 11, 2001 victims by the U.S. federal government was about $1.7 million. Three typical payments were summarized. A 26-year-old woman who worked as an accountant for annual compensation of $50,000 at a financial services company in the World Trade Center received $1.6 million. She was single and lived with her mother. A 40-year-old New York City firefighter, whose annual compensation was $71,300, received $1.5 million. He was single and was survived by two parents. A 33-year-old man who worked as an equities trader, for annual compensation of $2 million, at Cantor Fitzgerald in the World

Trade Center received an undisclosed sum. No projected awards were released for people who made more than $231,000 a year. He was married, with two children (Chen, 2004). Account for the differences between the compensation awards in 2004 and the NHTSA human life estimate of $200,725 in 1972.

4. Could Ford engineers have banded together and postponed the market release of the Pinto?

5. Did Ford engineers meet their professional responsibilities of protection of public safety, technical competence, and timely communication of negative and positive results to management?

Chapter 4

1981: Kansas City Hyatt Regency Skywalk Collapse

THE REPORTED STORY

The *New York Times* Abstract:

> The death toll rose to 111 in the Hyatt Regency Hotel accident today, as officials began trying to determine what caused the collapse of two walkways suspended above the hotel lobby. (Stuart, 1981)

THE BACK STORY

KANSAS CITY HYATT REGENCY HOTEL DESIGN

The Kansas City Hyatt Regency Hotel was designed as a hotel with three main sections: a 40-story tower section, a function block, and a connecting atrium area. The atrium was a large open area approximately 36 m by 44 m and 15 m high. Three suspended walkways spanned the atrium at the second, third, and fourth floors (Figure 4.1). These walkways connected the tower section to the function block. As shown in Figure 4.1, the third-floor walkway was independently suspended from the atrium roof trusses. In contrast, the second-floor walkway was suspended through six connections from the fourth-floor walkway, which was suspended from the roof framing (Pfrang, 1982).

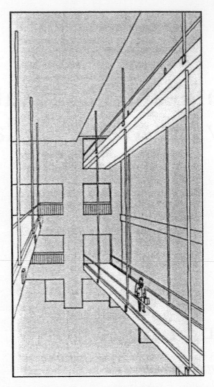

Figure 4.1 Schematics of the second-, third-, and fourth-floor walkways, looking south. From Pfrang, 1982. Republished with permission of ASCE.

HYATT PROJECT HIERARCHY

Hotel development began in 1976. Design and construction were conducted by specialized teams, under the direction of PBNDML Architects. The design team consisted of the architect, mechanical engineer, electrical engineer, and structural engineer. After the owner, Crown Center Redevelopment, chose the architect, the architect then chose the rest of the design team. The design team received a fixed fee for services rendered. The construction team consisted of the general contractor, Eldridge Construction Co., and its subcontractors, which included the structural steel fabricator and erector, Havens Steel Co. Havens subcontracted detailing work to WRW Engineering. The general contractor was chosen by the owner by its bid for the contract; the subcontractors were chosen by the general contractor by their bids.

The structural engineer, GCE International, was represented on this project by Daniel Duncan, a project engineer in charge of the actual structural engineering work. Duncan worked under the direct supervision of GCE President Jack Gillum. Though not bound by direct contracts, the structural engineer had certain control and authority over the construction team members through contract documents. No portion of the Hyatt project could commence until the shop drawings for that work had been approved by the structural engineer.

In developing structural steel aspects of a building like the Hyatt project, the structural engineer may design and analyze steel-to-steel members and connections. Calculations are performed to determine the strength and adequacy of the connection to carry the loads for which it is designed. In the corresponding structural drawings, these design details are called out as special "section details" within the structural drawings. If a section detail is not included in the structural drawing for a particular connection, the fabricator receiving the drawings employs its steel detailer to choose an applicable connection from the American Institute for Steel Construction (AISC) Manual of Steel Construction. The steel detailer translates structural drawings into shop and erection drawings for use in construction by the fabricator's construction crew. Completed structural drawings are sealed with the personal seal of the licensed professional engineer who prepared the drawings or under whose direction and supervision such drawings were prepared. The seal is the equivalent of the engineer's signature and indicates his acceptance of responsibility for the design shown.

After shop and erection drawings are prepared by the steel detailer, a steel checker reviews them. The checker only checks the exact work of the detailer. The structural engineer then reviews the shop and erection drawings by the fabricator and stamps them with the engineering firm's review stamp. The stamp represents drawing "conformance with the design concept and compliance with the information given in the contract documents" (Administrative Hearing Commission, 1985).

ORIGINAL BOX BEAM HANGER ROD DESIGN AND MODIFICATIONS

In 1978, Jack Duncan designed the box beams and hanger rods that were the structural steel members supporting the second-, third-, and fourth-floor walkways. On the second- and fourth-floor walkways, $1\frac{1}{4}$-inch diameter round steel rods were intended to run from the ceiling down to

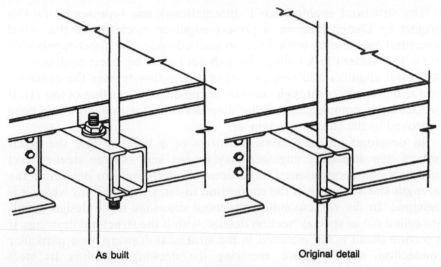

As built   **Original detail**

Figure 4.2 Hanger rod details: original and as built.
From Pfrang, 1982. Republished with permission of ASCE.

and through the fourth-floor box beams and were to continue down to and through the second-floor box beams, where the rods would terminate (Figure 4.2).

Duncan prepared a final section detail drawing, S405.1, that, at sections 10 and 11, depicted a box beam hanger rod connection typical of all such connections in the walkways. A nut and washer were illustrated, but no stiffeners or bearing plates were provided for added strength. The words "full development" and a weld symbol appeared at section 10. No final calculations for the loads associated with these connections were found in the project file where calculations were kept. After Gillum checked the final structural drawings for "design content and consistency with good engineering practice," he affixed his personal seal. The structural drawings were then sent to Havens Steel for preparation of shop and erection drawings.

Havens Steel subcontracted to WRW because Havens had too many projects. Specifically, head engineer William Richey subcontracted detailing work to Ken Warner, principal at WRW. WRW prepared 42 structural shop and erection drawings. For the weld at section 10, Warner selected a typical minimum assembly weld.

Havens was also responsible for purchasing the steel to be used while the drawings were being prepared. When Havens buyer Carl Bennett could only find shorter lengths than the 46 feet required for the steel rods, he informed Richey, who then had the length change communicated to

WRW. WRW modified the connections in shop drawing 30 and erection drawing E-3 to show this new double rod arrangement of 31'3" and 15'11" lengths for the second- and fourth-floor walkways, respectively.

Duncan approved the change of offsetting the rods at the fourth-floor box beam connections (see Figure 4.2). When one of the PBNDML architects called Duncan about the safety of this change to two rods, the architect was assured by Duncan that the change did not affect the structural integrity of the system. Although Duncan later testified that he performed a web shear calculation after his conversation with the architect, the calculation was not found in the project file. Because this was a "fast track" project, review of the shop and erection drawings was expedited in 10, rather than the typical 14, days. In February 1979, Duncan reviewed these drawings without making further calculations, and then applied the GCE stamp (Administrative Hearing Commission, 1985).

ATRIUM ROOF COLLAPSE

On October 14, 1979, part of the atrium roof collapsed during construction. GCE conducted an investigation of the collapse and determined that improper installation of a steel-to-concrete connection and inadequate provision for expansion resulting from faulty workmanship caused the collapse. The hotel owner also had an independent investigation conducted by structural engineering firm Seiden and Page that reached the same conclusions.

The owner and architect had directed GCE to check the design of all the steel, including steel-to-steel and steel-to-concrete, connections in the atrium. However, Gillum instructed Gregory Luth, who was an employee of GCE, to limit his design check to all structural members comprising the atrium roof. Duncan believed that Luth was to do a design check of all the atrium steel (Administrative Hearing Commission, 1985).

WALKWAY INVESTIGATION

When the hotel opened in 1980, it became a very popular nightspot, especially on Fridays, when an orchestra played for tea dance contests reminiscent of the 1940s. A year later, during a tea dance on July 17, 1981, the second- and fourth-floor walkways collapsed, leaving 114 people dead and 185 injured.

Soon thereafter, the mayor of Kansas City requested the National Bureau of Standards (NBS) to conduct an independent investigation of the collapse. NBS determined that the walkways began to fail when the bottom longitudinal welds near the ends of the fourth-floor box beams fractured and the bottom flanges deformed sufficiently to permit the box beam to slip down over the nut and washer at the lower end of the fourth floor to ceiling hanger rods. Because the second floor walkway was suspended from the fourth-floor walkway, loss of support for the fourth-floor walkway also caused the second-floor walkway to collapse.

By weighing selected sections of walkway debris, analyzing tape of the second-floor walkway collapse, and re-creating walkway parts for laboratory testing, NBS estimated the capacity of the actual fourth-floor box beam–hanger rod connections. The estimated mean capacities of the six connections ranged from 81 to 86 kN. However, each mean capacity was exceeded by the sum of the estimated dead load and upper-bound live load at each connection during the tea dance. Note that Kansas City Building Code required each connection to support forces imposed by combined dead and live load forces. For this type of connection, the Code also required an ultimate load capacity of 302 kN. Thus each fourth-floor connection was a candidate for initiation of walkway collapse.

Had the change in hanger rod detail not been made, the connections would still not have met the Kansas City Building Code. In terms of ultimate load capacity, the minimum value for this type of connection should have been 1.67 times 90 kN, or 151 kN. Based on test results, the mean ultimate capacity of a single-rod connection would have been approximately 91 kN, depending on the weld area (Pfrang, 1982).

ADMINISTRATIVE HEARING ACTIONS

On February 3, 1984, the Missouri Board of Architects, Professional Engineers and Land Surveyors filed a complaint against Daniel Duncan, Jack Gillum, and GCE International, charging gross negligence, incompetence, misconduct, and unprofessional conduct in the practice of engineering in connection with their performance of engineering services in the design and construction of the Hyatt Regency Hotel.

Duncan was found guilty of gross negligence in his preparation and completion of structural drawing S405.1, sections 10 and 11, and review of shop and erection drawings. Duncan was also found guilty of misconduct in his misrepresentation to the architects of the engineering acceptability of the double-hanger rod box beam connection. Gillum was found guilty of gross negligence in taking full personal and professional responsibility for all engineering design work performed, and for failing to review or ensuring someone reviewed structural drawing S405.1, sections 10 and 11, before placing his engineering seal. Gillum was also found guilty of unprofessional conduct in his lack of responsibility for all structural design aspects of the project. Further, he was found guilty of misconduct for his project engineer designee, Duncan, and for failure to review the atrium design. GCE International was found guilty of gross negligence, misconduct, and unprofessional conduct.

As disciplinary actions, Duncan and Gillum lost their licenses to practice engineering in the state of Missouri, while GCE had its certificate of authority as an engineering firm revoked (Administrative Hearing Commission, 1985).

APPLICABLE REGULATIONS

The Kansas City Building Code of 1978 could not be obtained from public records.

AN ENGINEERING PERSPECTIVE

It was the typical practice of GCE during shop and erection drawing review to have a technician check all the sizes and materials of structural members for conformance to design drawings, and to have the project engineer check engineering aspects of the drawings, including design work on connections where necessary. When technician Ed Jantosik conducted his portion of the review, he questioned project engineer Duncan about the strength of the rods called out on the shop drawings and the change from one rod to two. Duncan stated to Jantosik that the change to two rods was "basically the same as the one rod concept."

It should be noted that if Gregory Luth had been instructed to inspect all of the atrium, and not just its roof, Luth would have discovered flaws in the design of the second- and fourth-floor walkways (Administrative Hearing Commission, 1985).

REFERENCES

Administrative Hearing Commission, State of Missouri, Missouri Board for Architects, Professional Engineers and Land Surveyors vs. Daniel M. Duncan, Jack D. Gillum, and G.C.E. International, Inc. Case No. AR840239. Statement of the Case, Findings of Fact, Conclusions of Law, and Decision rendered by Judge James B. Deutsch, November 14, 1985.

BBC News, Paris inquiry spotlights concrete. *BBC News*, July 6, 2004. http://news.bbc.co.uk/1/hi/world/europe/3869437.stm.

Pfrang, E. O. and Marshall, R., Collapse of the Kansas City Hyatt Regency walkways. *Civil Engineering ASCE*, July 1982, 52, 65–68.

Stuart, R., Toll at 111 in Kansas City hotel disaster. *NY Times*, A1, July 19, 1981.
Wyatt, C., Paris terminal "showed movement." *BBC News*, May 26, 2004. http://news.bbc.co.uk/1/
    hi/world/europe/3751263.stm.

QUESTIONS FOR DISCUSSION

1. Should the steel fabricator and detailer assume more responsibility for their work on shop and erection drawings?

2. Download AISC's *Designing with Structural Steel: a Guide for Architects* at http://www.aisc.org/Content/ContentGroups/Documents/ePubs_Architects_ Guide/ArchitectsGuide.pdf. Read Part I, Basic Structural Engineering, on pages 21–36. Which elements were used in the Hyatt atrium design?

3. The AISC 2000 *Code of Standard Practice for Steel Buildings and Bridges* is contained within the appendix of this guide. Read pages 275–289, which describe procedures for design, shop, and erection drawings. How have these procedures been influenced by the Hyatt disaster?

4. During a fast-track project, the actual construction of a building begins before the design work is completed. In this way, the owner may avoid the full impact of escalating construction costs during the period of design and construction. Time pressure is put on the structural engineer to expedite shop drawing review, as the construction team is ready to proceed and lacks only the contractually required review and approval of shop drawings by the engineer (Administrative Hearing Commission, 1985). How ethical is the fast-track project delivery system?

5. On May 23, 2004, a 30-meter section of the roof of Terminal 2E of the Paris airport collapsed. This new terminal had opened only 11 months prior to the collapse and had been built using steel, concrete, and 36,000 sq m of reinforced glass. It had cost $900 million. Internationally renowned French architect Paul Andreu did not believe his futuristic design was to blame. He has created more than 50 airports around the world.

   Immediately after the collapse, construction details began to emerge. Trade unions in France claimed that builders were put under pressure to open the terminal on time. Airport cleaners admitted that two major water pipes had burst in the weeks before the accident. After the first water leaks, the cleaners had seen dust and particles falling from the ceiling. Airport officials confirmed that during an early stage of construction cracks appeared in the pillars holding up the concrete structure in an area of the terminal that did not collapse, which then had to be strengthened with carbon fiber. Shortly before the building's completion, 300 extra metal beams had been added to increase its stability (Wyatt, 2004).

On July 5, investigators announced that the metal support structure had perforated the concrete, causing it to split and collapse. Although the exact reasons were not known, the concrete was probably deteriorating (BBC News, 2004). How can this type of collapse be prevented in the future?

Chapter 5

1986: Challenger Space Shuttle Explosion

THE REPORTED STORY

The *New York Times* Abstract:

Cape Canaveral, FL, January 28—The space shuttle Challenger exploded in a ball of fire shortly after it left the launching pad today, and all seven astronauts on board were lost. (Broad, 1986)

THE BACK STORY

THE SPACE SHUTTLE DESIGN

The concept of a completely reusable space shuttle was first discussed in the 1960s, before the Apollo lunar landing spacecraft had flown. Over time, to minimize cost, the National Aeronautics and Space Administration (NASA) compromised on a reusable orbiter, an expendable external fuel tank carrying liquid propellants for the orbiters' engines, and two recoverable solid rocket boosters (Figure 5.1).

To provide for the broadest possible spectrum of civil and military missions, the shuttle was designed to deliver 65,000 lbs of payload to an easterly low-Earth orbit or 32,000 lbs to polar orbit. In early 1972, NASA estimated it would cost $6.2 billion to develop and test this three-part

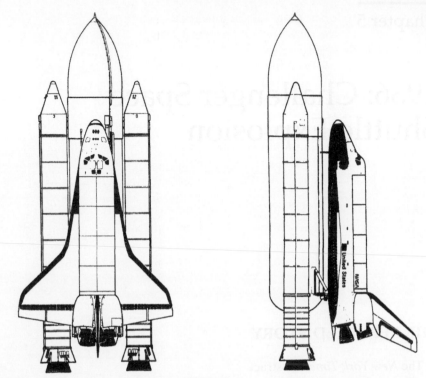

Figure 5.1   Two views of the space shuttle system: the orbiter, the expendable external fuel tank, and two recoverable solid rocket boosters.
Reprinted from Rogers Commission, 1986.

system. NASA awarded the contract for development of the orbiter and its main engines to Rockwell International Corporation, the contract for development of the external tank to Martin Marietta Denver Aerospace, and the contract for development of the solid rocket boosters to Morton Thiokol Corporation. Four space shuttle systems were built: the Columbia, the Discovery, the Atlantis, and the Challenger.

The orbiter is as large as a midsize airline transport and is constructed of an aluminum alloy skin stiffened with stringers to form a shell over frames and bulkheads of aluminum or aluminum alloy. The major structural sections are the forward fuselage, which encompasses the pressurized crew compartment; the mid fuselage, which contains the payload bay; the payload bay doors; the aft fuselage, from which the main engine nozzles project; and the vertical tail, which splits open along the trailing edge to provide a speed brake used during entry and landing. The payload bay is designed to securely hold a wide range of objects, from one or more communications satellites to be launched from orbit to cargo disposed on special pallets. It can carry 16 tons of cargo back from space.

To make it reusable, the orbiter needs to be protected from the searing heat generated by friction with the atmosphere when the craft returns to Earth. Temperatures during entry may rise as high as 2750° F on the leading edge of the wing and 600° F on the upper fuselage. The thermal protection system devised for the orbiter must prevent the temperature of the aluminum skin from rising above 350° during ascent or reentry. A carbon composite, consisting of layers of graphite cloth in a carbon matrix, protects the craft's nose cap and the leading edges of the wings. High-temperature ceramic tiles, about 6 inches square and varying in thickness from 1 to 5 inches, shield the areas subjected to the next greatest heat. Low-temperature tiles of the same material, designed to withstand 1200° F, shield areas requiring less protection.

Within the orbiter, the three high-performance rocket engines fire for approximately $8\frac{1}{2}$ minutes of flight after liftoff. At sea level, each engine generates 375,000 lbs of thrust at 100% throttle.

The propellants for the engines are the 143,000 gallons of liquid hydrogen fuel and 383,000 gallons of liquid oxygen oxidizer carried in the external tank. Built as a welded aluminum alloy cylinder with an ogive nose and a hemispherical tail, the external tank is 154 feet long and $27\frac{1}{2}$ feet in diameter. An intertank structure connects the two internal propellant tanks. A multilayered thermal coating covers the outside of the tank to protect it from extreme temperature variations during prelaunch, launch, and the first $8\frac{1}{2}$ minutes of flight. This insulation reduces the boil-off rate of the propellants, which must be kept at very low temperatures to remain liquid, and minimizes ice that might form from condensation on the tank exterior.

Initially, a wishbone attachment beneath the crew compartment connects the forward end of the orbiter to the external tank. A "bipod" also attaches the top of the orbiter to the external tank. About $8\frac{1}{2}$ minutes after liftoff, a command from the orbiter computer jettisons the external tank 18 seconds after the main engine cutoff. The tank breaks up upon atmospheric entry, falling into the planned area of the Indian or Pacific Ocean approximately an hour after liftoff.

The two solid-propellant rocket boosters are almost as long as the external tank and are attached to each side of it. They contribute about 80% of the total thrust at liftoff; the rest comes from the orbiters' three main engines. Roughly 2 minutes after liftoff and 24 miles down range, when the solid rockets have exhausted their fuel, explosives separate the boosters from the external tank.

Each solid rocket booster is made up of several subassemblies: the nose cone, solid rocket motor, and nozzle assembly. Each motor case is made of 11 individual cylindrical weld-free steel sections about 12 feet in diameter. The eleven sections are joined by tang-and-clevis joints held together by 177 steel pins around the circumference of each joint. A key joint is the solid rocket motor aft field joint, which connects the motor to the solid propellant (Figure 5.2).

Joint sealing is provided by two rubber O-rings, which are installed during motor assembly. Zinc chromate putty within the joint is intended to act as a thermal barrier to prevent direct contact of combustion gas with the O-rings. The O-rings are intended to be actuated and sealed by combustion gas pressure, displacing the putty in the space between the motor segments. This pressure-actuated sealing is required to occur very early during the solid rocket motor ignition transient, because the gap between the tang and clevis (the main field joints) increases as pressure loads are applied to the joint during ignition. If pressure actuation is delayed to the extent that the gap in the joint has opened considerably, it is possible that the rockets' combustion gases will blow by the O-ring and damage or destroy the seals.

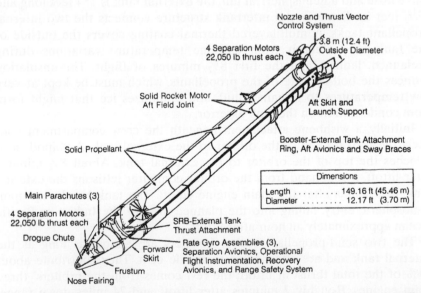

Figure 5.2 Cutaway view of the solid rocket booster showing solid rocket motor propellant and the aft field joint.
Reprinted from Rogers Commission, 1986.

EARLY PROBLEMS

The cost-plus-award-fee contract to design and build the shuttle solid rocket boosters, estimated to be worth $800 million, was awarded to Morton Thiokol on November 20, 1973. Thiokol's competitors for this contract were Aeroject Solid Propulsion Co., Lockheed Propulsion Co., and United Technologies. NASA's source evaluation board on the proposals rated Thiokol fourth under the design, development, and verification factor; second under the manufacturing, refurbishment, and product support factor; first under the management factor; and first under the cost factor. Because cost was the board's primary concern, Thiokol's lowest-cost bid won the contract.

Thiokol based its design primarily on that of the Air Force's Titan III solid rocket, one of the most reliable ever produced. However, Thiokol modified the joints so that the O-ring would take the brunt of the combustion pressure, with no other gas barriers present except an insulating putty. It also added a second O-ring to provide a backup in case the primary seal failed. Further, to simplify manufacture, the original joint seal design submitted in the proposal was modified from a face/bore seal to a double-bore seal. Whereas the Titan O-rings were molded in one piece, the Thiokol O-rings were made from sections of rubber O-ring material glued together.

Thiokol began testing the solid rocket motor in the mid 1970s. During an early important test in 1977, it was discovered that joint rotation (joint opening of up to 0.052 inch, rather than closing) occurred in the milliseconds after ignition. Although Thiokol engineers did not believe this would cause significant problems, NASA conducted further analysis and wrote several memos in 1977–1978 stating that the seal design was unacceptable. Further static motor tests conducted by NASA in July 1978 and April 1980 continued to demonstrate joint rotation, but Thiokol questioned the validity of these joint rotation measurements. In 1980, NASA empanelled a space shuttle verification/certification committee to study the flight worthiness of the entire shuttle system. Among other things, the committee recommended that NASA verify the field joint integrity, including firing motors at a mean bulk propellant temperature range of 40–90° F. The joint design passed these tests, and the solid rocket motor was certified on September 15, 1980. The solid rocket booster joint was classified as criticality category 1R, meaning that it contained redundant hardware (R = second O-ring), total element failure of which could cause loss of life or vehicle.

Although the O-rings were not degraded during the first space shuttle flight on April 12–14, 1981, primary O-ring erosion was discovered after the second flight in 1981. It occurred in the right solid rocket booster's aft field joint and was caused by hot motor gases. Although not present in every flight thereafter, O-ring erosion was discovered in 14 of 25 flights during

1983–1986. Blow-by of combustion gases past the O-ring was discovered in 10 of 25 flights during 1983–1986. Thiokol established an O-ring task force on August 20, 1985 to investigate the solid rocket motor case and nozzle joints and recommend both short-term and long-term solutions.

LAUNCH DELAYS AND SUBSEQUENT LAUNCH

In the 1980s, each space shuttle flight was designated by two numbers and a letter, such as 41-B. The first digit indicated the fiscal year of the scheduled launch, such as "4" for 1984. The second digit identified the launch site, with "1" signifying the Kennedy Space Center in Merritt Island, Florida, and "2" signifying Vandenberg Air Force Base in California. The letter corresponded to the alphabetical sequence for the fiscal year, with B being the second scheduled mission.

Mission 51-L of the Challenger was originally scheduled for July 1985, but by the time the crew was assigned in January 1985, launch had been postponed to late November to accommodate changes in payloads. The assigned crew included five career astronauts; one payload specialist from Hughes Aircraft, who would perform experiments that would support satellite redesign; and one payload specialist, Christa McAuliffe, who was to become the first teacher in space.

Launch of 51-L was further postponed three times and cancelled once. The first postponement established the launch date as January 23, 1986, in order to accommodate the final integrated simulation schedule that resulted from the slip in the launch date of mission 61-C. The second postponement established the launch date as January 26, primarily because of Kennedy work requirements produced by the late launch of mission 61-C. The third postponement established the launch date as January 27, due to unacceptable weather that was forecasted through the launch window. On January 27 the launch was cancelled when winds at the Kennedy runway increased and exceeded the allowable velocity for crosswinds.

For January 28, the weather was forecast to be clear and very cold, with temperature dropping into the low twenties overnight. During early morning hours of January 28, an ice inspection team examined accumulated ice at the launch pad twice, waiting until enough ice melted on the pad to recommend launch. At 11:38:00.010 A.M. Eastern Standard Time (EST), the Challenger began its final flight. The ambient air temperature at launch was 36° F, measured at ground level approximately 1000 feet from the 51-L launch pad. From liftoff until the space shuttle signal was lost 73 seconds later, no flight controller observed any indication of a problem. However, after 73 seconds, the Challenger exploded. All seven crew members perished.

PRESIDENTIAL COMMISSION INVESTIGATION

President Reagan appointed an independent commission to investigate the accident. Led by former Secretary of State William Rogers, the commission's mandate was to determine the probable causes of the accident and to develop recommendations for corrective action. Using mission data, subsequently completed tests and analyses, and recovered wreckage, the commission investigated all possible causes of the accident and determined that sabotage did not contribute to the accident.

Based on launch photographs and telemetry and tracking data, the commission ruled that the external tank, orbiter main engines, orbiter, payload/orbiter interfaces, and payload did not contribute to the cause of the accident. However, although the left solid rocket booster and all the components of right solid rocket boosters did not contribute to the accident, the right solid rocket motor was the principal cause.

During liftoff, a combustion gas leak through the right solid rocket motor aft field joint occurred. This leak was due to the reduced O-ring resiliency that resulted from the cold temperature. The ambient launch temperature was 36° F, which was 15° lower than the next coldest previous launch. The initial smoke after liftoff at 0.678 seconds came from the 270 to 310° sector of the aft field joint circumference, which faced the external tank. At 58.778 seconds into the flight, a small flame appeared in this joint area, growing into a continuous, well-defined plume. At 64.660 seconds, the swirling flame breached the external tank, when its shape and color abruptly changed as it began to mix with leaking hydrogen from the external tank. At 73.124 seconds, the entire aft dome of the external tank dropped away, releasing massive amounts of liquid hydrogen from the tank upward into the intertank structure. Within milliseconds, the external tank exploded, engulfing the orbiter in flames (Rogers Commission, 1986).

The commission concluded that neither Morton Thiokol nor NASA responded adequately to internal warnings about the faulty seal design. The specific conclusions included the following:

1. The joint test and certification program was inadequate. There was no requirement to configure the qualifications tests motor as it would be in flight, and the motors were static tested in a horizontal position, not in the vertical flight position.
2. Prior to the accident, neither NASA nor Thiokol fully understood the mechanism by which the joint sealing action took place.
3. NASA and Thiokol accepted escalating risk apparently because they "got away with it last time. . . ."
4. NASA's system for tracking anomalies for Flight Readiness Reviews failed in that, despite a history of persistent O-ring erosion and blow-by, flight was still permitted. . . .

5. The O-ring erosion history presented to Level I (National Space Transportation System director) at NASA headquarters in August 1985 was sufficiently detailed to require corrective action prior to the next flight.
6. A careful analysis of the flight history of the O-ring performance would have revealed the correlation of O-ring damage and low temperature. Neither NASA nor Thiokol carried out such an analysis; consequently, they were unprepared to properly evaluate the risks of launching the 51-L mission in conditions more extreme than they had encountered before. (Rogers Commission, 1986)

COMMISSION RECOMMENDATIONS

In order to ensure the return to safe shuttle flights, the presidential commission made the following recommendations:

1. The faulty solid rocket motor joint and seal must be changed, with an independent design oversight committee implementing the commission's design recommendations and overseeing this design effort.
2. The shuttle management structure should be reviewed. The program manager's responsibility should be redefined. A new safety advisory panel should report directly to the program manager.
3. NASA and the primary shuttle contractors should review all Criticality 1, 1R, 2, and 2R items and hazard analyses.
4. NASA should establish an Office of Safety, Reliability and Quality Assurance, reporting directly to the NASA Administrator. It would have direct authority for safety, reliability, and quality assurance throughout the agency.
5. NASA should take energetic steps to eliminate the non-communication tendency at Marshall Space Flight Center, whether by changes of personnel, organization, indoctrination or all three.
6. NASA must take actions to improve landing safety.
7. NASA must make all efforts to provide a crew escape system for use during controlled gliding flight and increase the range of flight conductions under which an emergency runway landing can be successfully conducted.
8. NASA must establish a flight rate that is consistent with its resources.
9. NASA should establish a system of analyzing and reporting performance trends of shuttle items designated Criticality 1. (Rogers Commission, 1986)

One year later, NASA submitted a report detailing all the actions taken to meet these recommendations (NASA, 1987). The first shuttle flight after the Challenger explosion was Discovery's launch on September 29, 1988.

APPLICABLE REGULATIONS

The space shuttle program is the major segment of NASA's National Space Transportation System (NSTS). It is divided into four levels:

Level I: NSTS director, responsible for the overall program requirements, budgets, and schedules.

Level II: NST program manager, responsible for shuttle program baseline and requirements. Provides technical oversight on behalf of Level I.

Level III: Program managers for orbiter, solid rocket booster, external tank, and space shuttle main engine, responsible for development, testing, and delivery of hardware to launch site.

Level IV: Contractors for shuttle elements, responsible for the design and production of hardware.

Prior to 1983, Level III was required to report all problems, trends, and problem closeout actions to Level II, unless a hardware-associated problem was not flight critical. However, a control board directive, submitted by Martin Raines, director of safety, reliability and quality assurance at Johnson, and signed by Level II on March 7, 1983, reduced the scope of reportable problems to Level II. The revised scope included only those problems that dealt with common hardware items or physical interface elements, and eliminated reporting on flight safety problems, flight schedule problems, and problem trends. According to a memo Mr. Raines wrote to the commission during its investigation in 1986, the documentation change was made in an attempt to streamline the system, because the old requirements were not productive for the operational phase of the shuttle program (Rogers Commission, 1986).

This document, Space Shuttle Program Requirements Control Board Directive 501152A, could not be obtained from public records.

AN ENGINEERING PERSPECTIVE

Thiokol engineer Roger Boisjoly inspected hardware from Flight 51C after it returned. This flight had been launched on January 14, 1985, during the coldest ambient temperature to date. Boisjoly found that hot combustion gases had blown by the primary seals on two field joints and had produced large arc lengths of blackened grease between the primary and secondary seals. Based on this discovery, he hypothesized that the low ambient temperature (resulting in a 53° F O-ring temperature) prior to launch caused reduced O-ring resiliency, which was responsible for the excessive blow-by. Further O-ring bench testing confirmed his hypothesis. Boisjoly became part of the Thiokol O-ring investigation task force formed in August 1985 (Boisjoly, 1987).

The night before the launch of Flight 51-L, because of Thiokol's concern that the launch would occur at temperatures in the low twenties, Thiokol scheduled a teleconference between Thiokol and NASA shuttle personnel at the Kennedy Space and Marshall Space Flight Centers. During this teleconference, Boisjoly presented his O-ring data, and Thiokol management recommended that the launch not occur until O-ring temperature reached at least 53° F, which was the lowest temperature of any previous flight. George Hardy, the Deputy Director of Science and Engineering at Marshall, was reported to have been "appalled" by Thiokol's recommendation (Rogers Commission, 1986) but could not launch over the contractor's objection (Boisjoly, 1987). Immediately thereafter, while NASA asked for a private caucus, Thiokol managers attempted to make a list of data to support a launch decision. The Thiokol engineers witnessed Senior Vice President Jerry Mason ask Vice President of Engineering Bob Lund to "take off his engineer hat and to put on his management hat" (Rogers Commission, 1986; Boisjoly, 1987). When the three groups reconvened their teleconference, Thiokol stated that although temperature effects were a concern, data were inconclusive, so a launch was recommended.

The temperature data of flights with O-ring incidents, which were presented by Boisjoly at the teleconference, are shown in Figure 5.3. When *all* flight data are added to the plot, including flights with no erosion or blow-by (see Figure 5.3), it becomes clear that reduced O-ring resiliency had occurred in every flight associated with a joint temperature less than 65° F (Rogers Commission, 1986).

After Boisjoly testified at the shuttle presidential commission, he experienced a hostile work environment at Thiokol. He was given an extended sick leave and then long-term disability for 2 years (Boisjoly, 1987). In 1988, Boisjoly received the American Association for the Advancement of Science's Scientific Freedom and Responsibility award "for his exemplary and repeated efforts to fulfill his professional responsibilities as an engineer by alerting others to life-threatening design problems of the Challenger space shuttle and for steadfastly recommending against the tragic launch of January 1986" (AAAS, 2005).

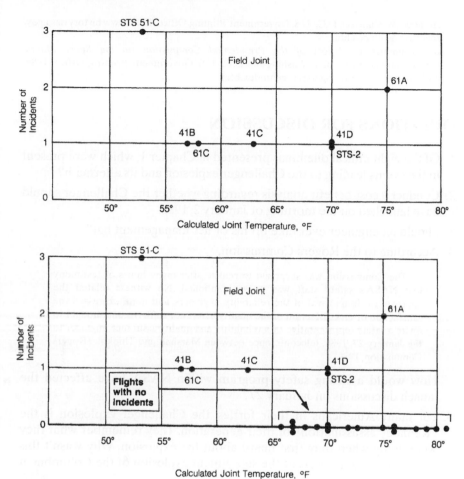

Figure 5.3 Plots of flights with and without incidences of O-ring thermal distress. Reprinted from Rogers Commission, 1986.

REFERENCES

American Association for the Advancement of Science (AAAS), *AAAS Science in Freedom and Responsibility Award 1988*. 2005, http://archives.aaas.org/people.php?p_id=331.

Boisjoly, R. M., *Ethical Decisions—Morton Thiokol and the Space Shuttle Challenger Disaster.* 1987, www.onlineethics.org/essays/shuttle/index.html.

Broad, W. J., Thousands watch a rain of debris; shuttle explodes, killing all 7 aboard. *NY Times,* A1, January 29, 1986.

National Aeronautics Space Administration, *NASA Report to the President: Implementation of the Recommendations of the Presidential Commission on the Space Shuttle Challenger*

Accident. Washington, D.C.: U.S. Government Printing Office, 1987. www.history.nasa.gov/
rogersrep/genindex.htm.

Rogers Commission, *Report of the Presidential Commission on the Space Shuttle
Challenger Accident,* vol 1. Washington, D.C.: U.S. Government Printing Office, 1986.
www.history.nasa.gov/rogersrep/genindex.htm.

QUESTIONS FOR DISCUSSION

1. Of the eight ethical dilemmas presented in Chapter 1, which were present in the events leading to the Challenger explosion and its aftermath?

2. Conduct a cost-benefit analysis regarding whether the Challenger should have launched on the morning of January 2, 1986.

3. Should an engineer ever take off his or her management hat?

4. According to the Rogers Commission,

> The Commission was surprised to realize after many hours of testimony that NASA's safety staff was never mentioned. No witness related the approval or disapproval of the reliability engineers, and none expressed the satisfaction or dissatisfaction of the quality assurance staff. No one thought to invite a safety representative or a reliability and quality assurance engineer to the January 27, 1986, teleconference between Marshall and Thiokol. (Rogers Commission, 1986)

   How would a strong safety program within NASA have affected the launch discussions on January 27?

5. For many Americans in their forties, the Challenger explosion is the "Kennedy assassination" of their generation. They remember what they were doing when they first heard about the explosion. Why wasn't this disaster enough to prevent the subsequent explosion of the Columbia in 2003?

Chapter 6

1989: Exxon Valdez Oil Spill

THE REPORTED STORY

The *New York Times* Abstract:

A tanker filled to capacity with crude oil ran aground and ruptured yesterday 25 miles from the southern end of the Trans-Alaska Pipeline, spewing her cargo into water rich in marine life. (Shabecoff, 1989)

THE BACK STORY

THE TRANS-ALASKA PIPELINE SYSTEM

After oil was discovered in Prudhoe Bay on the northern coast of Alaska in 1968, the Alyeska Pipeline Service Company was formed by the owner companies: BP Exploration, ARCO, Exxon, Mobil, Amerada Hess, Phillips, and Union. Alyeska determined that the most economic method of transporting oil from Prudhoe Bay to the U.S. west coast was oil transport through a pipeline from the bay to Valdez, followed by oil tanker transport south. President Richard Nixon signed the Trans-Alaska Pipeline Authorization Act on November 16, 1973.

The Trans-Alaska Pipeline System (TAPS) consists of an extensive 800 mile pipeline (Figure 6.1), 11 pump stations, and an oil terminal at Valdez; it cost more than $8 billion to build (USDIBLM, 2005).

Figure 6.1   A section of the Trans-Alaska Pipeline that is elevated to prevent permafrost from melting.
Courtesy U.S. Army Corps of Engineers.

TAPS is fed by several North Slope fields, including the Prudhoe Bay Oil Field. Prudhoe accounts for one-fourth of total domestic U.S. production and, through 1996, about one-eighth of U.S. consumption. The Valdez terminal contains 18 holding tanks, each of which holds about 0.5 million barrels of crude oil. Smaller storage facilities at Valdez add another 0.2 million barrel capacity. The average flow from the North Slope drilling sites for many years was 1.8 million barrels per day, or more than 650 million barrels per year. The Valdez shipping lanes through Prince William Sound are shown in Figure 6.2.

Alaska is very dependent on TAPS. Between 1969 and 1987, Alaskan state taxes amounted to $1.5 billion per year; federal taxes amounted to $2 billion per year. During this same period, Alyeska made about $2.4 billion per year in profit. Every year, each Alaska resident receives a TAPS dividend check of between $800 and $1000. This conflict of interest may explain why Alaskans allowed Alyeska to let construction and operations requirements, which were conditions of congressional TAPS approval, lapse. Even though the traditional Valdez industry is fishing, with more successful fishermen able to bring in annual incomes of six figures, Alyeska was allowed to extensively pollute Valdez waters.

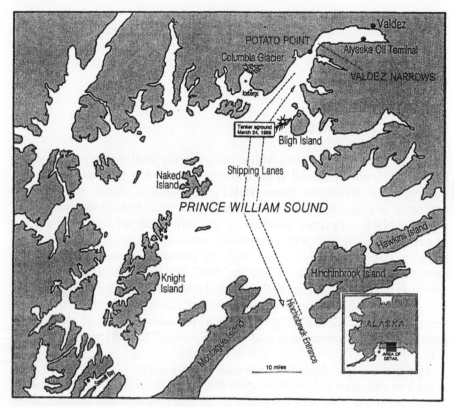

Figure 6.2 Prince William Sound and its shipping lanes.
Reprinted from Skinner, 1989, with additional citations.

These unfulfilled congressional requirements included 14 additional storage tanks, an incinerator to destroy toxic sludge produced by the terminals' operations; and stainless steel (replaced by less expensive carbon steel) in the toxic vapor recovery system of the storage tanks. The carbon steel pipe sprang dozens of leaks, resulting in more toxic vapors being released into the atmosphere. Although Alyeska had promised that its oil fleet would be composed of double-hulled tankers, almost all tankers reaching Valdez were single hulled.

The state of Alaska estimated that up to 1000 tons of hydrocarbons per week entered the air through vents on the decks of the tanks. However, Alyeska argued that tanker emissions were not its responsibility. Ballast water also polluted the environment. Ballast water is carried by tankers traveling to the Alyeska terminal to remain stable. The water is stored in the same tanks that will be filled with oil and is pumped out before oil is loaded. Although Alyeska was obligated to clean ballast water in its ballast treatment plant before discharge into Valdez waters, untreated ballast water and sludge from holding tanks were both regularly discharged (Keeble, 1999).

OIL SPILL PREPAREDNESS

At the time of the Exxon Valdez oil spill, six contingency plans were in place to coordinate oil spill response efforts. On the national level, the National Response Team (NRT) provided national support for response actions related to oil discharges and hazardous substance releases. NRT supported emergency responders at all levels by means of technical expertise and equipment, assisted in the development of training, coordinated responses with neighboring countries, and managed the National Response System. NRT actions were to be primarily conducted through Regional Response Teams (RRT). The RRT plan for Prince William Sound outlined a system for mechanical oil removal as a primary spill response strategy, which included chemical dispersant preauthorization procedures and wildlife protection guidelines for an oil spill.

Under state law, Alyeska developed a contingency plan specifically for rapid and effective responses to spills from vessels in trade with Alyeska's Valdez terminal. This plan gave priority to containment and cleanup of oil spills to prevent or minimize the amount of oil reaching 136 sensitive areas around Prince William Sound. The plan covered scenarios for three spill sizes, including one for an 8.4 million gallon spill in which approximately 50% of oil would be recovered at sea either directly after the spill or at a later time. The Exxon shipping Company Headquarters Casualty Response Plan, which was a voluntary document not required by federal law or regulation, defined the organization and responsibilities of a casualty management team and headquarters oil spill assistance team but had no specific details of actions to be conducted.

The captain of the Port Prince William Sound Pollution Action plan implemented provisions of the national and regional plans, taking into account the Alyeska plan. The state of Alaska plan listed the U.S. Coast Guard as having "basic investigative and enforcement responsibilities for oil spills that occur on coastal waters bordering Alaska" (Skinner, 1989).

These theoretical contingency plans were not rigorously implemented. According to Jerry Nebel, a former oil spill coordinator for Alyeska, "We knew exactly what was coming, where we were supposed to be, and we still messed it up. [Oil spill] drills were a farce, comic opera." In March 1988, Alyeska conducted an inventory of cleanup equipment but found only half of the emergency lights. The other half was being readied for use in Valdez's winter carnival. Other components listed in the Alyeska contingency plan that were missing included half the required length of 6-inch hose; 3700 feet of boom (a floating fence that is a staple of oil cleanup), amounting to 15% of what was required; and 8 of 10 blinking barricades (Keeble, 1999).

Four years before the Exxon Valdez ran aground, Captain James Woodle, the Alyeska Valdez port commander, wrote a confidential letter to management stating that "due to a reduction in manning, age of equipment, limited training and lack of personnel, serious doubt exists that [we] would be able to contain and clean up effectively a medium or large size oil spill" (Palast, 2003).

THE LAST VOYAGE OF THE EXXON VALDEZ

The Exxon Valdez arrived at Valdez and docked at the Alyeska Terminal at 10:48 P.M. on Wednesday, March 22, 1989, under the command of Captain Joseph Hazelwood. The ship measured 166 feet at the beam and 987 feet from bow to stern. The next morning, after spending the night on board, the crew discharged ballast water and filled 12 cargo tanks with 1,286,738 barrels (54,042,996 gallons) of crude oil. When the Exxon Valdez began its voyage to the west coast at 9:12 P.M. on March 23, the entire crew was exhausted. If the congressionally mandated requirement that officers have 6 hours' off-duty time within the 12-hour period prior to departure had been followed, the ship could not have sailed that evening.

As it moved from Valdez Narrows to the Valdez Arm, the ship encountered heavy ice, which stretched in a cone shape clear across the shipping lanes from Point Freemantle to within 0.9 miles of Bligh Reef. Rather than slow the ship down to go through the ice, Hazelwood opted to go around where the ice was thinnest. The ship that had just preceded the Exxon Valdez out of the channel, the Arco Juneau, had successfully executed this maneuver and avoided Bligh Reef. However, through a series of rudder turns, during which time Hazelwood was not always on the bridge, the Exxon Valdez ran aground on Bligh Reef around 12:20 A.M. on March 24. Hazelwood ordered the engine shut down and notified the Marine Safety Office at 12:28 that his vessel had struck the reef.

Later media reports overemphasized Hazelwood's alcoholism. Although it is true that Hazelwood violated Coast Guard policy by drinking liquor less than 4 hours before taking command of a vessel, charges of drunkenness and misconduct against him were dismissed at his 1990 administrative hearing before Coast Guard officials. More likely, fatigue was a major factor in the rudder turn missteps of the crew (Keeble, 1999).

OIL SPILL CLEANUP

Based on the implementation quality of the oil spill preparedness plans detailed earlier, it is not surprising that a huge oil spill would not be quickly recovered. Exxon reported that 10.8 million gallons of oil spilled into Prince William Sound; however, further examination of oil taken by other Exxon vessels immediately out of the area and of the intensity of oil effects suggest the spill was probably 23 to 25 million gallons. Pumping oil from the Valdez's tanks and reballasting the ship simultaneously took 11 days. Alyeska did not reach the Valdez for 15 hours; when it did reach the scene, it did not have sufficient equipment. Eighteen hours into the response, no boom had been deployed around the tanker or slick and only two small skimmers were operating. A skimmer is a mechanical device that removes oil from water. For the next few days, Alyeska maintained a token presence at the spill but simultaneously declined to respond to calls from local fishermen ready to volunteer dozens of boats for oil recovery.

Moreover, the state, Exxon, and the Coast Guard could not decide whether or not to use dispersants. When properly applied within 24 hours, dispersants cause an oil slick to fracture into tiny droplets that sink down into the water column, but also to rapidly release hydrocarbons that are extremely toxic. According to its plan, Alyeska should have recovered 100,000 barrels of oil by the end of the third day; instead, it recovered 3000 barrels. The dispersant discussion became moot at the end of 3 days, as the sea changed from docile to torturous. Twenty-five-foot waves draped oil onto rocks, as planes were grounded and boats confined to harbors.

By the evening of April 7, 15 days into the spill, the oil slick occupied an area of 18 square miles and had traveled 180 miles. Following the first storms, much of the spilled oil formed a mousse composed of 55% water. The mousse was impossible to burn off the surface and could not be easily skimmed. Further, conversion to mousse significantly impeded evaporation, dissolution, and microbial degradation. Emulsified tar balls that began to appear on shorelines on the Kenai Peninsula and Kodiak Island tended to retain toxic compounds at serious levels. By May 18, 56 days after the spill, the oil slick had traveled 470 miles (Keeble, 1999).

Approximately 1300 miles of shoreline were contaminated. Exxon implemented a variety of cleanup techniques for the different shorelines (e.g., mixed sand and gravel beaches, exposed tide flats) exposed to oil. Exxon reported that $2.1 billion was spent on cleanup over four summers. At its peak, the cleanup effort included 10,000 workers, about 1000 boats, and roughly 100 airplanes and helicopters. However, it is widely believed that wave action from winter storms did more to clean the beaches than all the human effort involved (EVOSTC, 2005).

It is estimated that 250,000 seabirds, 250 bald eagles, 2800 sea otters, 300 harbor seals, up to 22 killer whales, and billions of salmon and herring eggs died as a direct result of the oil spill. Seventy-four percent of these birds were murres, which will need many years to recover to former numbers. The eagles that died had ingested contaminated prey; their population seems to be rebounding. Although the otter population in the eastern, largely unoiled portion of the sound continues to increase at normal rates, the population in the heavily oiled western portion remains low. Possibly the sea otters there continue to suffer "subtle and difficult to detect" spill-related effects, including chronic liver and kidney damage. Harbor seals had been in a state of decline in the sound prior to the oil spill. After a sharp drop in population in 1989 and 1990, their previous trend of decline of 6% per year has resumed in both the oiled and unoiled areas. The majority of the killer whales that died came from one pod; the other five pods that currently use the sound for feeding and social activity have maintained normal population levels. Salmon and herring juveniles released from local hatcheries are now more susceptible to predation because of decreased growth rate from oil exposure and concomitant decrease in vitality (Keeble, 1999).

OIL SPILL INVESTIGATION

At the request of President George Bush, the National Response Team investigated the causes of the Exxon Valdez oil spill. An abbreviated version of report recommendations are given here:

1. Prevention is the first line of defense. We must continue to take steps to minimize the probability of oil spills.
2. Preparedness must be strengthened. Exxon, Alyeska, the state of Alaska, and the federal government were all unprepared for an oil spill of this magnitude.
3. Response capabilities must be enhanced to reduce environmental risk. Both public and private research are needed to improve

cleanup technology, as oil spills are difficult to clean and recovery rates are low.

4. Some oil spills may be inevitable. This awareness makes it imperative that we work harder to establish environmental safeguards that reduce the risks associated with oil production and transportation.
5. Legislation on liability and compensation is needed.
6. The United States should ratify the International Maritime Organization 1984 Protocols. Expeditious ratification is essential to ensure international agreement on responsibilities associated with oil spills around the world.
7. Federal planning for oil spills must be improved.
8. Studies of the long-term environmental and health effects must be undertaken expeditiously and carefully (Skinner, 1989).

LAWSUITS

On February 27, 1990, a federal grand jury in Anchorage indicted Exxon and its shipping subsidiary on five criminal counts. Two felony charges were based on the 1972 Ports and Waterways Safety Act and the Dangerous Cargo Act; three misdemeanors were based on the Clean Water Act, the Refuse Act, and the Migratory Bird Act. A settlement among Alaska, the federal government, and Exxon was reached in October 8, 1991. Exxon entered guilty pleas for violating provisions of these acts, paid a fine of $150 million, and settled damage claims of $900 million. Of the $150 million fine, $125 million was forgiven to Exxon for the company's previous expenses and cooperation, thus bringing the total settlement to $1.025 billion (Keeble, 1999).

Additionally, on September 14, 1994, an Anchorage jury awarded $5 billion in punitive damages to hundreds of members of a class lawsuit against Exxon. Exxon appealed to delay payments for a decade. On January 24, 2004, federal judge Russell Holland directed ExxonMobil to pay $4.5 billion in punitive damages and approximately $2.25 billion in interest; Exxon is in the process of appealing this decision (DWT, 2004).

APPLICABLE REGULATIONS

The Exxon Valdez oil spill was prosecuted under general maritime law. As a direct result of the oil spill, Congress enacted the Oil Pollution Act of 1990. This Act appears in Title 33, Chapter 40, of the United States Code

(U.S.C.). The United States Code is the codification by subject matter of the general and permanent laws of the United States. Section 2702 (33U.S.C.§2702) of Chapter 40 is given below:

Title 33 Subchapter I—Oil Pollution Liability and Compensation
Sec. 2702—Elements of Liability

(a) In general
Notwithstanding any other provision or rule of law, and subject to the provisions of this Act, each responsible party for a vessel or a facility from which oil is discharged, or which poses the substantial threat of a discharge of oil, into or upon the navigable waters or adjoining shorelines or the exclusive economic zone is liable for the removal costs and damages specified in subsection (b) of this section that result from such incident.

(b) Covered removal costs and damages
(1) Removal costs
The removal costs referred to in subsection (a) of this section are—
(A) all removal costs incurred by the United States, a State, or an Indian tribe under subsection (c), (d), (e), or (l) of section 1321 of this title, under the Intervention on the High Seas Act (33 U.S.C. 1471 et seq.), or under State law; and
(B) any removal costs incurred by any person for acts taken by the person which are consistent with the National Contingency Plan.
(2) Damages
The damages referred to in subsection (a) of this section are the following:
(A) Natural resources
Damages for injury to, destruction of, loss of, or loss of use of, natural resources, including the reasonable costs of assessing the damage, which shall be recoverable by a United States trustee, a State trustee, an Indian tribe trustee, or a foreign trustee.
(B) Real or personal property
Damages for injury to, or economic losses resulting from destruction of, real or personal property, which shall be recoverable by a claimant who owns or leases that property.

      (C) Subsistence use
             Damages for loss of subsistence use of natural resources, which shall be recoverable by any claimant who so uses natural resources which have been injured, destroyed, or lost, without regard to the ownership or management of the resources.

      (D) Revenues
             Damages equal to the net loss of taxes, royalties, rents, fees, or net profit shares due to the injury, destruction, or loss of real property, personal property, or natural resources, which shall be recoverable by the Government of the United States, a State, or a political subdivision thereof.

      (E) Profits and earning capacity
             Damages equal to the loss of profits or impairment of earning capacity due to the injury, destruction, or loss of real property, personal property, or natural resources, which shall be recoverable by any claimant.

      (F) Public services
             Damages for net costs of providing increased or additional public services during or after removal activities, including protection from fire, safety, or health hazards, caused by a discharge of oil, which shall be recoverable by a State, or a political subdivision of a State.

(c) Excluded discharges
    This subchapter does not apply to any discharge—
    (1) permitted by a permit issued under Federal, State, or local law;
    (2) from a public vessel; or
    (3) from an onshore facility which is subject to the Trans-Alaska Pipeline Authorization Act (43 U.S.C. 1651 et seq.).

(d) Liability of third parties
    (1) In general
        (A) Third party treated as responsible party
            Except as provided in subparagraph (B), in any case in which a responsible party establishes that a discharge or threat of a discharge and the resulting removal costs and damages were caused solely by an act or omission of one or more third parties described in section 2703 (a)(3) of this title (or solely by such an act or omission in combination with an act of God or an act of war), the third party or parties shall be treated as the responsible

party or parties for purposes of determining liability
under this subchapter.

(B) Subrogation of responsible party

If the responsible party alleges that the discharge or
threat of a discharge was caused solely by an act or
omission of a third party, the responsible party—

   (i)   in accordance with section 2713 of this title, shall
pay removal costs and damages to any claimant;
and

  (ii)   shall be entitled by subrogation to all rights of the
United States Government and the claimant to
recover removal costs or damages from the third
party or the Fund paid under this subsection.

(2) Limitation applied

(A) Owner or operator of vessel or facility

If the act or omission of a third party that causes an
incident occurs in connection with a vessel or facility
owned or operated by the third party, the liability of the
third party shall be subject to the limits provided in
section 2704 of this title as applied with respect to the
vessel or facility.

(B) Other cases

In any other case, the liability of a third party or parties
shall not exceed the limitation which would have been
applicable to the responsible party of the vessel or facility
from which the discharge actually occurred if the
responsible party were liable. (USC, 2005)

AN ENGINEERING PERSPECTIVE

The original radar equipment at the Coast Guard vessel traffic control
center in Valdez enabled simultaneous 24-hour surveillance of both Bligh
Reef and Valdez Narrows. However, in 1984 this equipment, which was
manufactured by AIL/Eaton, was replaced by Raytheon radar. Civilian
radar technician Pat Levy maintained the Coast Guard's Valdez radar
equipment. He learned that the agency was planning this replacement in
an effort to save money, and disagreed with the decision because he did
not consider the new equipment to be as potent or reliable. His disagree-
ment led to writing his congressman, Don Young, on February 29, 1984
that "I still can't help feeling that this is . . . bringing an oil tanker disaster
in the Sound closer to a reality."

When Young relayed Levy's concerns to the Coast Guard, its comman-
dant, Admiral James Gracey, wrote back that the new radar would be as
good as the old, without compromising safety. A Coast Guard radar expert
made the same claim at the Exxon Valdez disaster hearing in 1989.
However, within a year of the debut of the Raytheon equipment, Coast
Guard Commander Michael Cavett complained about poor reception
during the bad weather common to the Sound and asked for an upgrade of
radar at Potato Point. Cavett wrote in April, 1985 that "the installation of a
10 cm radar system could improve tracking ability in rain, wind, and snow.
I request one of the 3 cm radar systems at Potato Point be replaced with a
10 cm system." Note that a shorter wavelength enables better definition of
the target on the radarscope, but for a shorter range. During precipitation,
3 cm radar is more attenuated (reduced) than is 10 cm radar. This 3 cm
system was still in use when the Exxon Valdez collided with Bligh Reef
(Jones, 1989).

REFERENCES

Banerjee, S., *Arctic National Wildlife Refuge: Seasons of Life and Land.* Seattle: Mountaineers
    Book, 2003.
Barrionuevo, A., A dirty little footnote to the energy bill. *NY Times,* C1, April 15, 2005.
Davis Wright Tremaine (DWT), *Federal Judge Directs ExxonMobil to Pay Punitive Damages &
    Interest to Those Harmed by Exxon Valdez Oil Spill.* Press Release, January 28, 2004.
Diamond, J., *Collapse: How Societies Choose to Fail or Succeed.* New York: Viking, 2005,
    329–357.
Exxon Valdez Oil Spill Trustee Council (EVOSTC), *Oil Spill Facts: Questions & Answers.*
    April 30, 2005. http://www.evostc.state.ak.us/facts/qanda.html.
Jones, S., Blueprint for disaster: Empty promises. *Anchorage Daily News,* A1, October 15,
    1989.
Keeble, J., *Out of the Channel: The Exxon Valdez Oil Spill in Prince William Sound.* Spokane,
    WA: Eastern Washington University Press, 1999.
Kennedy, R. F., Jr., *Crimes Against Nature: How George W. Bush and His Corporate Pals
    Are Plundering the Country and Hijacking Our Democracy.* New York: HarperCollins,
    2004, 108.
Kolbert, E., Wasted energy. *New Yorker,* April 18, 2005, 55–56.
Mickelson, B. J., *MTBE Contamination of Ground Water.* Internal Exxon Memorandum,
    August 23, 1984. http://www.ewg.org/reports_content/withknowledge/pdf/068_002.pdf.
Palast, G., Exxon lubricated by Bush judges. *The Observer,* B1, September 2, 2003.
Shabecoff, P., Exxon vessel hits reef, fouling water that is rich in marine life; largest U.S. tank
    spill spews oil off Alaska coast. *NY Times,* A1, March 25, 1989.
Skinner, S. K., et al., *The Exxon Valdez Oil Spill: A Report to the President.* Washington, D.C.: U.S.
    Government Printing Office, 1989. 68 pp. http://www.epa.gov/history/topics/valdez/04.htm.
United States Code (USC), 2005. www.straylight.law.cornell.edu/uscode.
U.S. Department of Interior, Bureau of Land Management (USDIBLM), *TAPS Guide.* 2005.
    http://tapseis.anl.gov/guide/index.cfm.

QUESTIONS FOR DISCUSSION

1. In other textbooks, environmental ethics is considered an ethical dilemma. Environmental ethics is the study of moral issues concerning the environment. Environmental ethics is not defined as a separate ethical dilemma in this text because it is considered a subset of "protection of public safety." Provide reasons for this subclassification.

2. As noted in the text, TAPS production for many years was 1.8 million barrels per day. In recent years, this has dropped to only 1 million barrels per day. The need for increased production to satisfy U.S. demand recently convinced the majority in the Senate in 2005 to approve a national budget that would open a 1.5 million acre coastal region of the Arctic National Wildlife Refuge to drilling. It is estimated that 7.5 billion barrels may be recoverable, with production beginning to flow about a decade after drilling begins (Kolbert, 2005). Meanwhile, this land near the North Slope oil fields is the calving ground for more than 100,000 caribou of the Porcupine herd (Banerjee, 2003). Further, if the fuel-efficiency standards for cars and light trucks that were implemented by Jimmy Carter in 1979 had continued through 1986 (Ronald Reagan relaxed the standards in 1986), the United States would no longer have needed Persian Gulf oil after 1986 (Kennedy, 2004). Conduct a cost-benefit analysis regarding whether drilling in the Arctic National Wildlife Refuge should occur.

3. Read Chapter 11 of *Collapse*, a book by Jared Diamond (Diamond, 2005). Diamond is a MacArthur Foundation Fellow and Pulitzer Prize winner for his previous book, *Guns, Germs, and Steel*. In this chapter, Diamond details the divergent economic fates of Haiti and the Dominican Republic, two countries that share the same island land mass in the Caribbean. What are principal reasons that Diamond believes Haiti, which was originally more prosperous than the Dominican Republic, is now a poorer country? What lessons does this chapter have for Alaskan state officials?

4. As this textbook goes to press in 2005, Enron court cases continue to make headlines and Congress must again decide whether to pass an energy bill that includes a waiver that would protect oil companies from all methyl tertiary butyl ether (MTBE) liability lawsuits filed since September 2003. This additive has been used in gasoline since 1990 as an oxygen enhancer to reduce auto carbon monoxide emissions. In its 2003 product safety bulletin, Lyondell Chemical, the largest MTBE manufacturer, stated that even less than one part per billion imparted a "distasteful odor and taste" to groundwater that could make it "unsuitable for consumption." MTBE has been detected in 1861 water systems

in 29 states, serving more than 45 million Americans (Barrionuevo, 2005). Is there a difference in ethics between the energy industry and other industries? Discuss.

5. Because MTBE dissolves easily in water and does not readily cling to soil, it contaminates groundwater supplies more readily than other gasoline components. An Exxon chemical engineer wrote the first of several internal memos in 1984, warning of MTBE's "ground water incident costs and adverse public exposure" (Mickelson, 1984). But because MTBE is a byproduct of gasoline refining and is readily available, the oil industry chose to use MTBE as an oxygenate when Congress mandated in 1990 that some form of oxygenate be added to gasoline (Barrionuevo, 2005). Should the manufacturers of MTBE be liable for MTBE cleanup costs?

Chapter 7

1989: San Francisco–Oakland Bay Bridge Earthquake Collapse

THE REPORTED STORY

The *New York Times* Abstract:

A devastating earthquake rocked the San Francisco Bay area at rush hour last night, killing at least 200 people, collapsing a mile-long span of an Interstate highway and wrecking part the Bay Bridge to Oakland. (Barron, 1989)

THE BACK STORY

TRANSPORTATION IN THE BAY AREA IN THE 1920S

During the 1920s, the San Francisco Bay area was the most densely settled region of California. As San Francisco and Oakland grew during this time, transportation between these two major cities became more important. Passenger and vehicular ferries carried citizens between both ports across the San Francisco Bay. In 1929, passenger ferries carried 36 million passengers and vehicular ferries carried 10 million passengers with their automobiles. During these years, the number of annual passengers was decreasing while the number of annual autos was increasing. This was not surprising, because Californians owned one car for every 2.7 people in 1929. East Bay residents used the East Bay Highway to travel from suburbs to the Oakland area. Vehicles were ferried from Berkeley to Hyde Pier in San Francisco. Once in San Francisco, vehicles used the Bay Shore Highway to travel between San Francisco and San Mateo.

The Southern Pacific railroad also had indirect routes between the two cities. One railroad crossing across the bay spanned Dumbarton Port in Alameda County and Redwood City in San Mateo County; the second crossing spanned Benicia and Martinez.

With this transportation system in place, the San Francisco–Oakland Bay bridge began its construction in 1933. Designed only to improve highway and mass transit services between the East Bay and San Francisco, this bridge was congested on the day of its dedication in 1936 and has remained congested almost without interruption (HAER, 1999).

BRIDGE DESIGN AND CONSTRUCTION

The Bay Bridge design was based on the following requirements:

- Capacity of six lanes for highway traffic and at least two operative and one passing or emergency track for interurban trains;
- One or two bridge spans between San Francisco and Yerba Buena island (the west channel), and one span between the island and Oakland;
- Vertical and horizontal clearances of 220 feet and 1650 feet, respectively, mandated by the War Department.

Unlike other prominent bridges built at the time, such as the Golden Gate Bridge, the Bay Bridge was designed by committee, rather than under the vision of one designer/architect. It was efficiently designed during a 24-month period, from early 1931 to early 1933. In 1931 the California legislature authorized the California Toll Bridge Authority (CTBA), which it had created in 1929, to build the bridge and provided $650,000 for its design. C. H. Purcell, who studied civil engineering at Stanford University, was appointed chief engineer for the bridge; Charles Andrew, who studied civil engineering at the University of Illinois, was named as his assistant, or bridge engineer. When Purcell and Andrew's requested exemption from civil service was approved, they hired a talented team of more than 50 engineers along with allied specialists, including surveyors, draftsmen, and a clerical staff. Two prominent team members were Ralph Modjeski, one of the best known bridge engineers of the early 20th century, and Daniel Moran, one of the best known foundation engineers.

Due to the horizontal clearances requested by the War Department, a suspension design, rather than a cantilever design, was selected for the west channel. A suspension bridge is formed when a roadway is suspended from vertical cables, which are in turn attached to two or more main cables. These main cables hang from two towers, and their ends are anchored in bedrock or concrete. Because a suspension bridge had never been built over such a long span (4100 feet), engineers carefully modeled and analyzed a design containing a central anchorage at the midpoint of a double suspension span (Figure 7.1a).

Other advantages of this design were the lower cost to build and aesthetics. Based on geologic studies, bedrock could be found at a depth of 300 feet in the west channel. The stiff clay overlaying this rock would provide the greatest frictional resistance to sinking. This depth limited each western bridge span to 1400 feet.

Because of the wider variation of foundation depth in the east channel, only the first of the East Bay crossing piers could be taken to bedrock. It was imperative that the weight across the east span be minimized to compensate for unstable abutments. Thus a cantilever design was selected for the east span, with steel columns substituting for masonry columns to reduce dead load and long distances planned between expansion joints (Figure 7.1b). A cantilever bridge is formed by self-supporting arms anchored at and projecting toward one another from the ends; they meet in the middle of the span, where they are connected together or support a third member (HAER, 1999). The western and eastern spans were to be connected by a 1700-foot tunnel through Yerba Buena Island, drilled through shale rock.

Construction began in 1933 and was completed 6 months ahead of schedule in 1936 (Figure 7.2).

The upper level of the bridge had three automobile lanes in each direction. The lower level had three lanes for trucks and buses and two standard-gauge electric railway tracks for interurban trains. Public bonds were sold to finance the bridge's $77.6 million cost. At the time of completion, the Bay Bridge set the record for the longest and largest bridge in the world (California State Parks, 2004).

RAILWAY RETROFIT

In 1955 the interurban railway through the bridge was abandoned. Because the total bridge traffic was far busier than anticipated (33 million annual vehicles versus projections of 11 million in 1950), each deck of the bridge was reconfigured for one-way traffic. This reconfiguration required strengthening of the upper deck to accommodate added truck and bus traffic, increasing the height of the upper-deck tunnel through Yerba Buena to accommodate truck traffic, and reconstructing the approaches on both sides of the bridge to accommodate unidirectional traffic patterns.

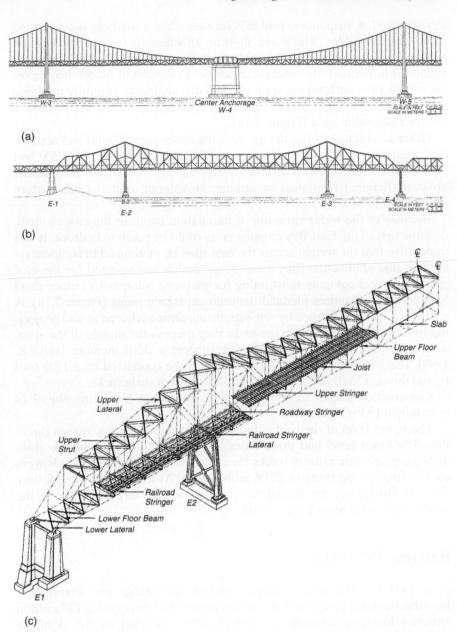

(a)

(b)

(c)

Figure 7.1   Bridge spans: (a) suspension spans, (b) cantilever truss spans, (c) cantilever upper deck detail.
Reprinted from HAER, 1999.

Figure 7.2   The San Francisco–Oakland Bay Bridge.
Courtesy U.S. Army Corps of Engineers.

Complicating this reconfiguration were the facts that the bridge could not be closed to traffic because it was one of the busiest stretches of road in the country, and that the work would be conducted in stages of many years in order to distribute the cost over many fiscal budgets.

Reconstruction occurred in three phases from 1959 to 1961. During the third phase, one of the most complex tasks was strengthening the upper deck of the East Bay crossing. In order to reduce the length of the longitudinal spans for the deck slabs, new joists were installed midway between existing joists, which were attached to existing stringers (Figure 7.1c). Additional stringers were bolted into the bottom flanges of the original stringers. The East Bay crossing was then resurfaced because the deck was showing fatigue after 27 years of use and tiles needed to be eliminated that had delineated six lanes of traffic. Reconstruction was completed on October 12, 1963.

LOMA PRIETA EARTHQUAKE

The reconstructed bridge was rebuilt to carry a capacity of 110,000 daily cars; by 1989 it was carrying 240,000 daily cars. Luckily, when the Loma Prieta earthquake occurred at 5:04 P.M. on October 17, 1989, afternoon rush traffic was lighter than usual because many commuters had changed their schedules to accommodate the third game of the World Series, which was to begin at 5:30 P.M. at Candlestick Park. Drivers on the suspension span felt side-to-side movement. Drivers in the tunnel felt no movement but were aware of problems because the tunnel lights were extinguished. Drivers on the East Bay crossing felt violent movement, especially through truss spans. The Loma Prieta earthquake measured 7.1 on the Richter scale.

Damage occurred at Pier E-9, which is the juncture of the through truss and the continuous deck truss sections of the East Bay crossing. The earthquake caused the through truss and continuous truss sections to move in opposite directions. It is estimated that the continuous truss moved 1 inch north and 5.5 inches east. As a result, both the upper and lower 50-foot connector sections fell, with the lower section affected by the upper section falling first. One driver died during this collapse. This section of the bridge was repaired within 1 month (HAER, 1999).

NEW BRIDGE

Because it was originally built in the 1930s, the existing Bay Bridge does not conform with current operational standards set by the American Association of Highway and Transportation Officials. It will also not be able to withstand another large earthquake. Geologists insist that an earthquake much larger than the Loma Prieta earthquake has a 70% likelihood of occurring in the next 30 years in the Bay Area. According to a Caltrans study completed in 1997, it is more cost effective to replace the East Bay crossing with a new bridge than to seismically retrofit the existing structure.

The new east crossing bridge design was based on the following requirements:

- New bridge alignment either north or south of current bridge, so the current bridge can remain open during construction;
- Capability of being reopened within 24 hours of a large earthquake to accommodate emergency response vehicles and heavy equipment (also known as the lifeline criteria);
- Any tower would not be of a height that overwhelms the existing west towers;
- Bridge would not have two decks;
- Ability to withstand an 8.1 to 8.2 magnitude earthquake as close as 3.2 miles away (Loma Prieta earthquake center was 30 miles away).

The 1998-approved design consists of an Oakland short low-rise span to the bridge, a 1.5-mile roadway, the world's first single-tower self-anchored suspension span, and a box-girder roadway connecting to Yerba Buena Island. During an earthquake, the roadway is designed to sway and slide, and the tower is designed to allow for movement (CAFJ, 2005). However, according to an Army Corps of Engineers study completed in 2000, the performance of this bridge design during an 8.1 to 8.2 magnitude earthquake as close as 3.2 miles away cannot be guaranteed because the design is not based on meeting this strict seismic safety requirement. Further, this design does not meet the lifeline criteria (USACOE, 2000).

Originally scheduled to open in December 2006, 2010 of the bridge's 11,525 feet had yet to be built as of March 2005. Total cost estimates have ballooned from $1.3 billion to more than $5 billion. Governor Schwarzenegger called for a simpler design eliminating the tower in the remaining 2010 feet to reduce costs (Chea, 2005). However, Bay Area legislators believe Schwarzenegger's idea would cost more and result in further delays because it requires design and environmental approvals. Additionally, the California legislature is divided regarding whether only Bay Area commuters, through tolls, or all Californians should pay for cost overruns of $2.9 billion, based on the original estimated cost by Caltrans in 2001. Meanwhile, the Federal Bureau of Investigation is checking allegations that the foundation of the concrete span built so far is riddled with shoddy welds. As of May 2005, work continues on the part of the bridge reaching Oakland but remains stalled on the last feet to Yerba Buena Island. Expected completion for the new bridge is now around 2011, 22 years after the Loma Prieta earthquake (Schmidt, 2005).

APPLICABLE REGULATIONS

According to the California Environmental Quality Act (CEQA), Ch. 1, §21000, of the California Code of Regulations:

§21000. Legislative intent
The Legislature finds and declares as follows:

(a) The maintenance of a quality environment for the people of this state now and in the future is a matter of statewide concern.
(b) It is necessary to provide a high-quality environment that at all times is healthful and pleasing to the senses and intellect of man.

(c) There is a need to understand the relationship between the maintenance of high-quality ecological systems and the general welfare of the people of the state, including their enjoyment of the natural resources of the state.

(d) The capacity of the environment is limited, and it is the intent of the Legislature that the government of the state take immediate steps to identify any critical thresholds for the health and safety of the people of the state and take all coordinated actions necessary to prevent such thresholds being reached.

(e) Every citizen has a responsibility to contribute to the preservation and enhancement of the environment.

(f) The interrelationship of policies and practices in the management of natural resources and waste disposal requires systematic and concerted efforts by public and private interests to enhance environmental quality and to control environmental pollution.

(g) It is the intent of the Legislature that all agencies of the state government which regulate activities of private individuals, corporations, and public agencies which are found to affect the quality of the environment, shall regulate such activities so that major consideration is given to preventing environmental damage, while providing a decent home and satisfying living environment for every Californian (State of California, 2005).

Certainly, earthquake safety is covered by this Act. However, in 1998, California Senate Bill 60 (Ch. 327) was passed, which included a provision that extended a previous CEQA exemption for seismic retrofit projects on state-owned toll bridges until the date the retrofit activities are certified complete or June 30, 2005, whichever comes first.

AN ENGINEERING PERSPECTIVE

The feasibility of building the San Francisco–Oakland Bay Bridge was first investigated by government engineers in 1924. They determined that this bridge was impractical due to earthquake faults and the difficulty of finding solid anchorage on the muddy bottom. However, President Herbert Hoover, who was an engineer, took an interest in the bridge idea. With California Governor C. C. Young, they appointed the Hoover-Young Commission. In its report submitted in August 1930, the commission stated that not only was the bridge necessary for area development, but that the bridge was "entirely feasible from economic and construction viewpoints"

(UC Berkeley Library, 1999). After building was completed in 1936, the first major earthquake (at least 6.0 in magnitude) that the bridge experienced was the Loma Prieta earthquake.

REFERENCES

Bai, M., Drip, drip, drip. *NY Times Sun Mag,* 78, June 8, 2003.
Barron, J., Utilities disrupted; thousands evacuated at World Series game—6.9 on Richter. *NY Times,* A1, October 18, 1989.
California Alliance for Jobs (CAFJ), *The Bridge: East Span Design Basics & Seismic Safety.* 2005. http://www.newbaybridge.org/the_bridge/east_span_design_basics.html.
California State Parks, *The San Francisco–Oakland Bay Bridge Photo Collection.* 2004. http://www.smrc.parks.ca.gov/index.php?option=com_content&task=view&id=55&Itemid=81
Chea, T., McClintock calls Bay Bridge project a "fiasco," *Sacramento Union,* A1, March 16, 2005.
Historic American Engineering Record (HAER), *San Francisco–Oakland Bay Bridge, Spanning San Francisco Bay.* HAER No. CA-32. Washington, D.C.: Library of Congress, 1999. 273 p. 20 drawings.
Schmidt, S., A bridge to nowhere, *San Diego Union-Tribune,* A1, May 8, 2005.
State of California, *California Environmental Quality Act.* 2005. http://www.co.el-dorado.ca.us/planning/ceqa.html.
U.S. Army Corps of Engineers (USACOE), *Final Report: Evaluation & Assessment of Proposed Alternatives to Retrofit/Replace the East Span of the San Francisco–Oakland Bay Bridge* October 27, 2000. http://www.oaklandbridge.com.
University of California at Berkeley Library, *San Francisco-Oakland Bay. Bridging the Bay: Bridging The Campus.* 1999. http://www.lib.berkeley.edu/news_events/exhibits/bridge/sfobay.html.

QUESTIONS FOR DISCUSSION

1. List missed opportunities for upgrading the stability of the San Francisco–Oakland Bay Bridge. Why was the bridge never stabilized to survive a major earthquake?

2. Read "Drip, Drip, Drip," which appeared in the *New York Times Sunday Magazine* in 2003 (Bai, 2003). State and city budgets must be balanced each year, whereas the federal budget is allowed to be in deficit. How does a federal deficit affect a state's budget?

3. Name some of the top budget issues for the state of California this year. How important is Bay Bridge stability compared with these other issues?

4. Throughout the 20th century, the aesthetic of the Golden Gate Bridge, which connects San Francisco to Sausalito, has been praised over that of the

original Bay Bridge. One of the protests to Governor Schwarzenegger's proposal is that the removal of the self-anchored suspension span will render the new bridge a "freeway on stilts." How important are aesthetics to the bridge design process?

5. Investigate the backgrounds and analytic process of the Engineering and Design Advisory Panel (EDAP) tasked in 1997 with selecting the design of the new Bay Bridge. Did the engineers on this panel adhere to the professional responsibilities described in Chapter 1? Did they engage in any of ethical dilemmas detailed in Chapter 1?

Chapter 8

1994: Bjork-Shiley Heart Valve Defect

THE REPORTED STORY

The *New York Times* Abstract:

A unit of Pfizer Inc. has agreed to pay $10.75 million to settle Justice Department claims that the company lied to get Federal approval for a mechanical heart valve that has fractured, killing hundreds of patients worldwide. (Meier, 1994)

THE BACK STORY

HEART VALVES

During the cardiac cycle, blood is transported from the atria to the ventricles, to the systemic and pulmonary circulation, and back again. Within the heart, blood flow is regulated by the four valves: the aortic, mitral, pulmonary, and tricuspid valves. Each valve is made up of a few thin folds of tissue, called leaflets or cusps, and keeps blood from flowing backward, or regurgitating, when closed.

A diseased valve may not open or close completely to regulate blood flow. Valvular disease may be congenital (present at birth) or may be caused by an infection that invades the tissue. It may also be caused by rheumatic fever, heart attack, stroke, or aging. Symptoms include dizziness, shortness of breath, fatigue, irregular heart rhythms, and fluid retention. The disease may be managed by drug therapy or, when severe, by surgical replacement.

A prosthetic replacement is necessary when stenosis or insufficiency occurs. During stenosis, the opening of the valve has decreased, causing the heart to work much harder to transport blood. During insufficiency, the valve is leaky, causing blood to flow backward. A prosthetic replacement may either be created from artificial materials or from animal or human cadaver tissue. Mechanical heart valves, made of artificial materials, last for life but may increase the risk of blood clots, causing a patient to take blood thinners for life. Biologic heart valves need to be replaced every 10 to 15 years.

THE BJORK-SHILEY HEART VALVE

Dr. Viking Bjork, a cardiac surgeon at the Karolinska Hospital in Stockholm, Sweden, first began to replace native valves in his patients with mechanical heart valves during the 1960s. After experience with the Bahnson Teflon cusp, Starr valve, and Kay-Shiley valve, he designed his own valve with American engineer Donald Shiley. They based their design on the tilting disc valve because this valve design decreased the pressure gradient across the valve compared with the gradient across the Kay-Shiley valve. Additionally, their free-floating disc valve, which was first composed of Delrin polyacetal plastic, opened in the aortic position to 60 degrees. The original flat disc was held in place by an inflow strut (shown in Figure 8.1 at top) and outflow strut, which were welded to the valve ring.

A carbon-coated double-flange universal sewing ring surrounded the valve. This first version of the Bjork-Shiley valve was first implanted in one of Bjork's patients in 1969. Later, in 1971, the flat disc material was changed from Delrin to Pyrolyte carbon.

In order to minimize blood clot formation after implantation, Bjork and Shiley modified the original disc shape to be convexo-concave in 1976. Additionally, the disc pivot point was moved 2.5 mm downstream, allowing the disc to pivot away slightly from the orifice ring in the open position. The inflow strut was made an integral part of the valve ring without any welds; the outflow strut was welded. The result was additional clearance between the disc and the flange, which diminished the low-flow area behind the disc by 50% and provided for a "washing effect" of the disc. It was claimed that this new shape not only decreased blood clot complications by 50% but kept the valve completely open with half the flow required with a flat disc. It also had a much more rapid reaction on closure, resulting in reduced

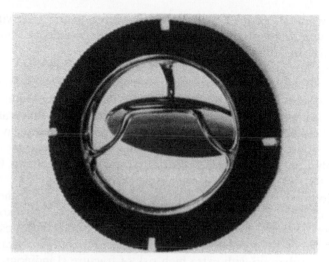

Figure 8.1   Bjork-Shiley monostrut convexo-concave heart valve.
Reprinted from Bjork, 1985, http://www.tandf.no/scj, by permission of Taylor and Francis.

regurgitation. Later, a second version of the convexo-concave valve was fabricated that included a modified opening angle of 70 degrees. This modified angle decreased the in vivo gradient by an average 15% across valves sized 21 to 33 mm, as compared with the gradient of the spherical disc valve.

To minimize the effect of manufacturing complications, Bjork and Shiley designed a monostrut valve in 1982, again based on an opening angle of 70 degrees. All portions of the monostrut valve became integral with the flange, as the valve was machined from a solid piece of metal and contained no welds. The cross-section of the outlet strut was 1.5 times greater than the area of the two wires of the convexo-concave valve (Bjork, 1984; Bjork, 1985).

BJORK-SHILEY VALVE COMPLICATIONS & INVESTIGATIONS

Eventually, 255,000 standard (Bjork, 1984) and 83,000 convexo-concave valves (Michaud, 1994), manufactured by Shiley Inc., were implanted worldwide. Over time, defects in implanted valves began to be reported. If the valve that failed was an aortic prosthesis, the patient generally died within 10 minutes. If the valve was a mitral prosthesis, the patient could survive for several hours (Lindblom, 1986).

As Bjork described in 1985, valve mechanical dysfunction was related to either obstruction of the disc movement or escape of the disc. Obstruction of disc movement occurred rarely and was traced to the skill with which the valve was implanted. However, the majority of defects were due to escape of the disc, disc rupture, or disc escape.

Disc rupture occurred infrequently. Somewhat more likely, the disc escaped if the outflow strut was dislocated during surgical insertion of the valve. However, the majority of defects were due to strut fracture (Bjork, 1985). In an implantation follow-up study of 3278 Bjork-Shiley valves over 15 years at the Karolinska Hospital, strut fracture did not occur in standard Delrin (n = 271), aortic standard Pyrolyte (n = 739), or monostrut (n = 377) valves. Only 1 out of 430 mitral standard Pyrolyte valves fractured. However, 6 out of 884 60-degree convexo-concave and 12 out of 577 70-degree convexo-concave valves fractured (Lindblom, 1986). In a further 3-year follow-up of these Karolinska patients, 2 more 60-degree convexo-concave and 9 more 70-degree convexo-concave valves fractured. These results indicated that the early production series of 70-degree convexo-concave valves constituted a subgroup with extra high risk of fracture (Lindblom, 1989). In another study of 24 explanted convexo-concave valves from 22 patients, 7 (29%) demonstrated a single-leg strut fracture and 2 demonstrated fatigue changes. 6 out of 7 of the single-leg strut fractures were welded by Shiley welder number 2295. Shiley, Inc. later admitted that 60-degree convexo-concave valves were modified to become the early production 70-degree convexo-concave valves (de Mol, 1994).

Based on structural analyses conducted by Shiley employees, stress loads on the struts were extremely low and insignificant, assuming normal properties of the fabrication materials. An internal investigation of manufacturing processes was conducted, with particular attention paid to the strut weld. Because clinical data had demonstrated that almost no outlet strut fractures had occurred in standard valves, comparisons were made between standard valve and convexo-concave valve welds.

As shown in Figure 8.2, the weld in the standard valve was consistent with homogeneous dendritic formations following the energy path from the welding torch.

However, all welds in the convexo-concave valve were not consistent. A small percentage from each weld lot demonstrated brittle areas within the weld, which resulted in phase segregation. Theoretically, cracks could form in these "brittle" areas in subsequent manufacturing steps or during valve implantation, and propagate during constant clinical cycling (Bjork, 1985). The observation of wear flats on the tips of outlet struts led to the conclusion that abnormally high loads were being applied during valve closure. These loads were due to changes in the outlet strut weld position and angle that allowed the disc the freedom to over-rotate on closure and sometimes make contact with the tip of the outlet strut, leading to very high bending loads. The

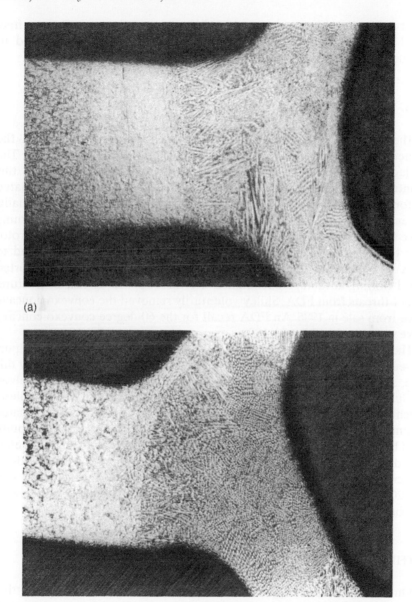

(a)

(b)

Figure 8.2 Photographs of a section through a weld: (a) perfect weld shows homogenous dendrite formation through the outflow strut in the drill hole and continuing 2 mm into the strut; (b) imperfect weld shows dendrite structure has not penetrated all of the outflow strut in the upper ring-strut angle, producing an anterior area of more brittle phase segregation. Reprinted from Bjork, 1985. http://www.tandf.no/scj, by permission of Taylor and Francis.

observation that there were no great differences in the wear flats on fractured and unfractured struts indicated that weld quality contributed to
fatigue fracture (Piehler, 2004).

U.S. GOVERNMENT INTERVENTION

The 60-degree convexo-concave valve was approved for sale in the
United States by the Food and Drug Administration (FDA) in 1979. The
70-degree convexo-concave valve was never approved for sale in this
country. In its premarket application to FDA, Shiley claimed an inflated
performance advantage of lowered blood clotting incidence. Additionally,
Shiley failed to provide FDA with all information it possessed concerning
valve fractures, including the fact that 18 valves had fractured during prototype testing. After the fracture problem became public, Shiley argued to
FDA that the valve should remain on the market because its purported
blood-clotting advantage outweighed the threat posed by risk of fracture.
Under threats from FDA, Shiley voluntarily removed the convexo-concave
valve from sale in 1986. An FDA recall for the 60-degree convexo-concave
valve was issued on May 23, 1990.

The United States filed a civil suit against Shiley and Pfizer, which purchased Shiley in 1979, under the False Claims Act and common law. By this
time, at least 250 patients had died from defective convexo-concave valves.
On July 1, 1994, Shiley agreed to pay the federal government $10.75 million,
which was the government estimate of payments for valves and valve-related
treatments through Medicare and the Veterans Health Administration, multi-
plied by two and a half times. This multiplication factor was an amount determined in lieu of penalties. Additionally, Shiley agreed to pay these federal
agencies for future medical costs related to the valves for an estimated total
of $20 million. As part of this settlement, Shiley admitted no liability
(Michaud, 1994).

OTHER LAWSUITS

One year before the government civil suit was settled, Shiley settled
with 259 valve recipients for an undisclosed amount, estimated to be worth
$26 million. A federal court appeal of a 1992 class-action settlement in
Cincinnati, which included all 83,000 Shiley heart valve recipients around
the world, was dismissed in 1994 (Michaud, 1994).

The settlement for this class-action suit, Arthur Ray Bowling, et al. v.
Pfizer Inc., et al., provided for a Consultation Fund of $80 million,

intended to provide claimants with funds to provide medical and psychologic consultation as they deemed best. It was to be divided equally among claimants after paying or providing for fees paid out of this fund. The settlement also provided $12.5 million for a Patient Benefit Fund, in order to conduct research on the diagnosis of strut fracture risk and to establish guidelines for diagnosis and valve replacement (Special Masters/Trustees, 1995). As of 2004, there had been 93 (71 foreign) qualified outlet strut fracture claims and 137 (55 foreign) qualified valve replacement surgery claims. The qualified valve replacement surgery claims included 38 qualified single leg fracture claims (Special Masters/Trustees, 2004).

APPLICABLE REGULATIONS

As a direct result of the Bjork-Shiley valve lawsuits, Congress enacted the Safe Medical Devices Act of 1990. This Act provides specific rules for medical device post-market surveillance. It appears in Title 21, Chapter 9, of the United States Code (U.S.C.). The General Rule under Section 360i (21U.S.C.§360i) is given below:

Title 21 Chapter 9 – Federal Food, Drug, and Cosmetic Act
Sec. 360i.—Records and reports on devices

(a) General rule
Every person who is a manufacturer or importer of a device intended for human use shall establish and maintain such records, make such reports, and provide such information, as the Secretary may by regulation reasonably require to assure that such device is not adulterated or misbranded and to otherwise assure its safety and effectiveness. Regulations prescribed under the preceding sentence—
(1) shall require a device manufacturer or importer to report to the Secretary whenever the manufacturer or importer receives or otherwise becomes aware of information that reasonably suggests that one of its marketed devices—
(A) may have caused or contributed to a death or serious injury, or
(B) has malfunctioned and that such device or a similar device marketed by the manufacturer or importer would be likely to cause or contribute to a death or serious injury if the malfunction were to recur;

(2) shall define the term "serious injury" to mean an injury that—
  (A) is life threatening,
  (B) results in permanent impairment of a body function or permanent damage to a body structure, or
  (C) necessitates medical or surgical intervention to preclude permanent impairment of a body function or permanent damage to a body structure;
(3) shall require reporting of other significant adverse device experiences as determined by the Secretary to be necessary to be reported;
(4) shall not impose requirements unduly burdensome to a device manufacturer or importer taking into account his cost of complying with such requirements and the need for the protection of the public health and the implementation of this chapter;
(5) which prescribe the procedure for making requests for reports or information shall require that each request made under such regulations for submission of a report or information to the Secretary state the reason or purpose for such request and identify to the fullest extent practicable such report or information;
(6) which require submission of a report or information to the Secretary shall state the reason or purpose for the submission of such report or information and identify to the fullest extent practicable such report or information;
(7) may not require that the identity of any patient be disclosed in records, reports, or information required under this subsection unless required for the medical welfare of an individual, to determine the safety or effectiveness of a device, or to verify a record, report, or information submitted under this chapter; and
(8) may not require a manufacturer or importer of a class I device to—
  (A) maintain for such a device records respecting information not in the possession of the manufacturer or importer, or
  (B) to submit for such a device to the Secretary any report or information—
    (i) not in the possession of the manufacturer or importer, or
    (ii) on a periodic basis,
    unless such report or information is necessary to determine if the device should be reclassified or if the device is adulterated or misbranded.

In prescribing such regulations, the Secretary shall have due regard for the professional ethics of the medical profession and the interests of patients. The prohibitions of paragraph (7) of this subsection continue to apply to records, reports, and information concerning any individual who has been a patient, irrespective of whether or when he ceases to be a patient. The Secretary shall by regulation require distributors to keep records and make such records available to the Secretary upon request. Paragraphs (4) and (8) apply to distributors to the same extent and in the same manner as such paragraphs apply to manufacturers and importers. (USC, 2005)

AN ENGINEERING PERSPECTIVE

In 1980, Bjork became aware that valve-strut failures were occurring and demanded corrective action. Although known as a surgeon, Bjork was in essence also a biomedical engineer who co-designed the Bjork-Shiley valve. Bjork threatened to publish cases of valve-strut failures. A panicked Shiley executive telexed: "ATTN PROF BJORK. WE WOULD PREFER THAT YOU DID NOT PUBLISH THE DATA RELATIVE TO STRUT FRAC-TURE." The executive's reason for holding off public exposure of the failure was "WE EXPECT A FEW MORE." Bjork kept silent (Palast, 1998) until he began publishing his corrective action investigations in 1985 (Bjork, 1985).

REFERENCES

Bjork, V. O., The development of the Bjork-Shiley artificial heart valve. *Clin Cardiol,* Jan, 1984, 7, 3–5.

Bjork, V. O., Metallurgic and design development in response to mechanical dysfunction of Bjork-Shiley heart valves. *Scand J Thorac Cardiovasc Surg,* Jan, 1985, 19, 1–12.

de Mol, B. A., Kallewaard, M., McLellan, R. B., van Herwerden, L. A., Defauw, J. J., and van der Graaf, Y., Single-leg strut fractures in explanted Bjork-Shiley valves. *Lancet,* Jan 1, 1994, 343, 9–12.

Food and Drug Administration (FDA), *Support Statement for Medical Devices: Current Good Manufacturing Practice (CGMP), Quality System (Q/S) Regulation.* 21 CFR Part 820, OMG No. 0910–0073. 2004. http://www.fda.gov/ohrms/dockets/98fr/04n-0034-ss00001.pdf.

Lindblom, D., Bjork, V. O., and Semb, B. K. H., Mechanical failure of the Bjork-Shiley valve. *J Thorac Cardiovasc Surg,* May, 1986, 92, 894–907.

Lindblom, D., Rodriguez, L., and Bjork, V. O., Mechanical failure of the Bjork-Shiley valve. *J Thorac Cardiovasc Surg,* Jan, 1989, 97, 95–97.

Meier, B., Pfizer unit to settle charges of lying about heart valve; Pfizer unit to settle U.S. charges. *NY Times,* A33, July 2, 1994.

Michaud, A., Shiley Inc. settles false-claims suit courts: Irvine-based firm and its parent to pay $10.75 million to U.S. in case involving flawed artificial heart valves. *LA Times,* A1, July 1, 1994.

Palast, G. The explosive truth behind U.S. wave of corporate crime. *London Observer,* A5, November 1, 1998.

Piehler, H. R., and Hughes, A. A., The role of postmarket surveillance in the medical device risk management system. In *Engineering Systems Symposium,* Cambridge, MA, March 31, 2004.

Special Masters/Trustees, *First Report of the Special Masters/Trustees Covering Period from February 28, 1992 to February 28, 1995.* Bowling-Pfizer Litigation. 1995. http://www.bowling-pfizer.com/images/reports/Report_1.pdf.

Special Masters/Trustees, *Twenty-First Report of the Special Masters/Trustees Covering Period from June 4, 2004 to November 4, 2004.* Bowling-Pfizer Litigation. 2004. http://www.bowling-pfizer.com/images/reports/twentyfirst_report.pdf.

United States Code (USC), 2005. http://www.straylight.law.cornell.edu/uscode.

United States Congress, *The Bjork-Shiley Heart Valve: "Earn While You Learn."* Committee Print 101-R, Subcommittee on Oversight and Investigations, Committee on Energy and Commerce, U.S. House of Representatives, February 1990.

QUESTIONS FOR DISCUSSION

1. Of the eight ethical dilemmas presented in Chapter 1, which were present in the events leading to the Bowling-Pfizer settlement?

2. Should Dr. Bjork have publicly disclosed in 1980 his request to Shiley for corrective action? Should he have disclosed this to FDA in 1980?

3. When the Safe Medical Device Act passed in 1990, FDA was given the power to recall medical devices. The number of Bjork-Shiley convexo-concave valves subject to strut failure was estimated to be fewer than 2000 per 89,000, or less than 2%. Should this device have been recalled? Discuss.

4. When the Safe Medical Device Act passed in 1990, medical device manufacturers were mandated to file Medical Device Reports (MDRs) with FDA when serious adverse device incidents occurred. An FDA review of the MDRs from 1991 revealed that 59 deaths and 929 serious injuries were attributed to design-related failures (FDA, 2004). No animal or clinical trials of the Bjork-Shiley convexo-concave design were conducted before valves of this design began to be implanted (Piehler, 2004). In fact, Congress dubbed the Shiley methodology of implanting its valves, then modifying its design iteratively (Delrin disc → Pyrolyte disc → 60-degree convexo-concave → 70-degree convexo-concave → monostrut) as "Earn While You Learn" (U.S. Congress, 1990). What types of testing should have been conducted before the first implant?

5. During discussions of congressional tort reform in early 2005, the Bowling v. Pfizer settlement was mentioned as an example of an excessive judgment. The legislation resulting from these discussions, the Class Action Fairness Act of 2005, limits attorney fees and moves jurisdiction from local to federal district courts (USC, 2005). Do you believe the Bowling v. Pfizer settlement was excessive? Discuss.

Chapter 9

1999: Y2K Software Conversion

THE REPORTED STORY

The *New York Times* Abstract:

Every major computer company is gearing up to cope with any problems that emerge as dates flip over to 2000 in millions of machines at midnight—first in Asia, then Europe, then the Americas; but IBM and Microsoft probably have the most at stake in terms of reputation, and potential liability; they have highest profile on issue, though for very different reasons; IBM introduced computing to corporations and governments in 1960's, and origins of Year 2000 problem—storage-saving convention of dropping first two numbers in dates of years—date back to mainframe era; and an estimated 70 percent of world's business data still resides on mainframe computers, most of them IBM machines; as dominant technology company of personal computer era, Microsoft is lightning rod for concern about Year 2000 problem partly because its Windows desktop is face of computing to most users. (Lohr, 1999)

THE BACK STORY

THE MILLENNIUM BUG

The millennium bug, also referred to as the year 2000 or Y2K problem, resulted from computer programming practices dating back to around 1960. When programmers used dates, they shortened years to a two-digit form to

109

save memory space. Although the cost of memory dropped substantially over the years, the convention of using the two-digit format (i.e., "60," rather than 1960) did not change until recently.

The rollover problem occurs when a computer program attempts to read the date "00," which could mean 1900 or 2000. If the year is read as "1900," unpredictable computer behavior may occur, which at worst could trigger a complete shutdown. Computers and other devices that are susceptible to this behavior were deployed in the utilities, health care, telecommunications, transportation, financial institutions, government, and general business sectors. The problem could occur in software, hardware, or data communication. The bug needed to be fixed for each individual case. Because the United States possesses one fourth of the world's computer assets, these fixes were not trivial.

SPECIAL COMMITTEE ON THE YEAR 2000 TECHNOLOGY PROBLEM

On April 2, 1998, the U.S. Senate unanimously voted to establish a new committee to address the year 2000 technology problem. Senator Bob Bennett was appointed chairman of this committee of seven senators. Bennett had been involved with the Y2K issue since assuming chairmanship of the Senate Banking Subcommittee on Financial Services and Technology the previous year. He had authored the Computer Remediation and Share Holder (CRASH) Protection Act, which required all publicly traded companies to fully disclose all details on their efforts to meet year 2000 readiness goals. On introduction of the Bennett bill, the Securities and Exchange Commission (SEC) issued a new legal bulletin requiring full disclosure by corporations.

This committee was created to study the impact of Y2K computer problems on the executive and judicial branches of the federal government, state governments, and private sector operations in the United States and abroad. After analysis, it would make recommendations for new legislation, amendments to existing laws, or administrative actions. Its existence was authorized through February 29, 2000 (United States Senate, 1998).

INTERIM ASSESSMENTS

After its establishment, the Committee held nine hearings by February 1999 on seven critical economic sectors: financial institutions, telecommunications, utilities, health care, transportation, government, and general business.

At this time, the financial services sector ranked ahead of nearly all other industries in its remediation and testing efforts. Legislation in Congress and action by the Committee had led to legal requirements on broker-dealers and publicly traded companies to disclose compliance information. By late 1998, the telecommunications industry had spent billions on Y2K fixes and was expected to have 99% of access lines in compliance by Fall 1999. Industry and government were working together to coordinate contingency plans in case there were failures. As a whole, the utility industry, composed of 3200 independent utilities, was configured to handle interruptions, blackouts, and natural disasters. However, local and regional outages remained a distinct possibility, depending upon the overall preparedness of the individual electric utility serving a given area.

In contrast, the nation's single largest industry, the health care industry, lagged significantly in preparedness. Generating $1.5 trillion annually, it is composed of 6000 hospitals, 800,000 doctors, 50,000 nursing homes, and hundreds of equipment manufacturers and suppliers and health care insurers. Because of limited resources and lack of awareness, rural and inner-city hospital had particularly high Y2K risk exposure. Sixty-four percent of hospitals had no plans to test their Y2K remediation efforts; 90% of physicians' offices were unaware of their Y2K exposure. Similarly, the nation's 670 domestic airports started Y2K compliance too late. The maritime shipping industry had not moved aggressively toward compliance. Public transit could be seriously disrupted. Several state and many local governments lagged in Y2K remediation, raising the risk of service disruption. Although the federal government had projected to spend more than $7.5 billion, it would not be able to renovate, test, and implement all of its mission-critical systems in time. However, wholesale failure of federal government services was not likely to occur.

In general, large companies had dealt well with the Y2K resources. Very small businesses were expected to survive using manual processes until Y2K remediation occurred. However, many small- and medium-sized businesses were extremely unprepared for Y2K disruptions (Senate, 1999a).

By the time the Committee published "The 100 Day Report" on September 22, 1999, much more progress had been made. The Health Care Financing Administration, the federal agency that oversaw Medicare payments, had made a nationwide effort to ensure its health claims payments system was Y2K compliant. The Federal Aviation Administration had successfully completed its effort to remediate national air traffic control systems. The Federal Emergency Management Administration (FEMA) was now engaged in national emergency planning in the event of major and minor Y2K disruptions. The heavily regulated insurance, investment services, and banking industries were farthest ahead of other industries in remediation.

Problems still remained. Some of the nation's 670 domestic airports remained at risk in areas such as jetway security systems and runway lighting. Likely there would be disruptions resulting in delays at some airports. Approximately 10 states were not prepared to deliver critical services such as unemployment insurance and other benefit payments. Approximately only 65% of state-critical systems were remediated as of May 1999, and only 25% of counties were ready as of June 1999. Health care, oil, education, agriculture, farming, food processing, and the construction industries all lagged behind other industries. The cost to regain lost operational capability for a mission-critical failure in these industries was estimated to range from $20,000 to $3.5 million, with an average of 3 to 15 days necessary to regain lost functions.

Early estimates for litigation because of Y2K-related failures ran as high as $1 trillion. To discourage litigation, the Y2K Readiness and Responsibility Act (Public Law No. 106-37) was passed to encourage remediation, not litigation, of Y2K problems (Senate, 1999b).

DAY ONE PREPARATION

"Day One Preparation" referred to the operational plans devised in governmental and private organizations to manage the January 1, 2000 date-change transition. It represented the largest simultaneous mobilization of resources in anticipation of a potential disaster or emergency. The governmental Y2K Information Coordination Center (ICC) established a broad communications and reporting network. Emergency operations centers were staffed by FEMA and emergency management agencies at municipal, county, and state levels. In private industry, Day One Preparation typically involved increased staffing of business and industry locations to both monitor and test key systems as the date rollover occurred, and to provide immediate on-premise technical assistance for any failures.

Several days before and after the date transition, ICC staff manned an operational desk and utilized the Information Collection and Reporting System (ICRS), which was ICC's primary data collection system. The ICRS allowed each user to review the status of each sector, including government services, finance, transportation, power and water utilities, telecommunications, health care, and business.

Y2K AFTERMATH

Although hundreds of computer problems were reported after the date transition, most were quickly corrected and none caused serious disruptions. Examples included Medicare payment delays, double-billing by some credit card companies, degradation of a spy satellite system, 911 problems in several locales, and a nuclear weapons plan system anomaly. However, no major problems were experienced in the United States or worldwide during the millennium date transition. This was an expected national result, because the United States spent an estimated $100 billion on Y2K. Additionally, information sharing, a focus on supplier interrelationships, and attention to contingency planning all contributed to a smooth transition. Further, Senate Committee hearings revealed a significantly lower failure rate than predicted for embedded chips. Analysis of testing during the last quarter of 1999 predicted an embedded chip failure rate of 0.001%, rather than the 2% to 3% failure rate projected in late 1998 and early 1999.

The Office of Management and Budget estimated that federal government Y2K spending reached $8.45 billion; the Commerce Department estimated that the U.S. government and businesses spent approximately $100 billion. A journalist from *Newsweek* estimated global Y2K spending of $500 billion (Senate, 2000).

Congress appropriated $3.35 billion in Y2K emergency supplemental funding. The Department of Defense received $1.1 billion; $2.25 billion went to nondefense departments and agencies. The top 10 nondefense agencies and departments receiving funds are depicted in Figure 9.1 (Senate, 2000).

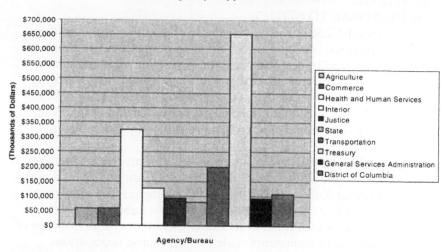

Figure 9.1 Y2K emergency supplemental funds for the top 10 nondefense agencies and departments.
From Senate, 2000.

APPLICABLE REGULATIONS

Section 101 of the Year 2000 Readiness and Responsibility Act from the U.S. Code (House Report 106-131) is given below:

SEC. 101. NOTICE PROCEDURES TO AVOID UNNECESSARY YEAR 2000 ACTIONS.

(a) NOTIFICATION PERIOD—Before filing a year 2000 action, except an action that seeks only injunctive relief, a prospective plaintiff shall send by certified mail to each prospective defendant a written notice that identifies, with particularity as to any year 2000 claim—

 (1) any symptoms of any material defect alleged to have caused harm or loss;

 (2) the harm or loss allegedly suffered by the prospective plaintiff;

 (3) the facts that lead the prospective plaintiff to hold such person responsible for both the defect and the injury;

 (4) the relief or action sought by the prospective plaintiff; and

 (5) the name, title, address, and telephone numbers of any individual who has authority to negotiate a resolution of the dispute on behalf of the prospective plaintiff.

Except as provided in subsection (c), the prospective plaintiff shall not commence an action in Federal or State court until the expiration of 90 days after the date on which such notice is received. Such 90-day period shall be excluded in the computation of any applicable statute of limitations.

(b) RESPONSE TO NOTICE

 (1) IN GENERAL—Not later than 30 days after receipt of the notice specified in subsection (a), each prospective defendant shall send by certified mail with return receipt requested to each prospective plaintiff a written statement acknowledging receipt of the notice and describing any actions it has taken or will take by not later than 60 days after the end of that 30-day period, to remedy the problem identified by the prospective plaintiff.

 (2) INADMISSIBILITY—A written statement required by this subsection is not admissible in evidence, under Rule 408 of the Federal Rules of Evidence or any analogous rule of evidence in any State, in any proceeding to prove liability for, or the invalidity of, a claim or its amount, or otherwise as evidence of conduct or statements made in compromise negotiations.

 (3) PRESUMPTIVE TIME OF RECEIPT—For purposes of paragraph (1), a notice under subsection (a) is presumed to be received 7 days after it was sent.

(c) FAILURE TO RESPOND—If a prospective defendant fails to respond to a notice provided pursuant to subsection (a) within the 30-day period specified in subsection (b) or does not describe the action, if any, that the prospective defendant has taken or will take to remedy the problem identified by the prospective plaintiff within the subsequent 60 days, the 90-day period specified in subsection (a) shall terminate at the end of that 30-day period as to that prospective defendant and the prospective plaintiff may thereafter commence its action against that prospective defendant.

(d) FAILURE TO PROVIDE NOTICE—If a defendant determines that a plaintiff has filed a year 2000 action without providing the notice specified in subsection (a) and without awaiting the expiration of the 90-day period specified in subsection (a), the defendant may treat the plaintiff's complaint as such a notice by so informing the court and the plaintiff in its initial response to the complaint. If any defendant elects to treat the complaint as such a notice—

  (1) the court shall stay all discovery in the action involving that defendant for the applicable time period provided in subsection (a) or (c), as the case may be, after filing of the complaint; and

  (2) the time for filing answers and all other pleadings shall be tolled during such applicable period.

(e) EFFECT OF CONTRACTUAL WAITING PERIODS—In cases in which a contract or a statute enacted before January 1, 1999, requires notice of nonperformance and provides for a period of delay prior to the initiation of suit for breach or repudiation of contract, the period of delay provided in the contract or the statute is controlling over the waiting period specified in subsections (a) and (d).

(f) SANCTION FOR FRIVOLOUS INVOCATION OF THE STAY PROVISION—In any action in which a defendant acts pursuant to subsection (d) to stay the action, and the court subsequently finds that the defendant's assertion that the suit is a year 2000 action was frivolous and made for the purpose of causing unnecessary delay, the court may award sanctions to opposing parties in accordance with the provisions of Rule 11 of the Federal Rules of Civil Procedure or the equivalent applicable State rule.

(g) COMPUTATION OF TIME—For purposes of this section, the rules regarding computation of time shall be governed by the applicable Federal or State rules of civil procedure.

(h) SPECIAL RULE FOR CLASS ACTIONS—For the purpose of applying this section to a year 2000 action that is maintained as a class action in Federal or State court, the requirements of the preceding subsections of this section apply only to named plaintiffs in the class action. (USC, 2005)

AN ENGINEERING PERSPECTIVE

Computer scientist Bob Bemer provided the first published warning about the millennium bug in a 1971 editorial for the *Honeywell Computer Journal* (Bemer, 1971). Known as the "father of ASCII," Bemer created ASCII in 1961 by assigning standard numeric values to letters, numbers, punctuation marks, and other characters. He also assisted Grace Hopper in creating the computer language Common Business Oriented Language (COBOL). In 2003 the Institute of Electrical and Electronic Engineers' Computer Society awarded him its Computer Pioneer medal.

Bemer's warnings were based on work he had conducted in the 1950s on genealogic records for the Church of Jesus Christ of Latter-Day Saints, when he realized that truncating a year's date was not worth the computer space saved. But Pentagon bureaucrats, among the largest computer users on Earth, refused to accept that 1999 was a better code than 99. The National Bureau of Standards agreed with the Pentagon, although it said programmers could voluntarily use four instead of two numbers.

Bemer began warning the public in 1971 (Sullivan, 2004). According to Dr. Fred Brooks, IBM's project manager for the IBM 360, the cost of using four-digit years decreased gradually as the wisdom of using them increased, with both lines crossing around 1970 (Weingarten, 1999). Bemer continued these warnings until he retired in 1982, even though the public reacted with derision (Sullivan, 2004).

REFERENCES

Bemer, R.W., What's the date? *Honeywell Computer J*, 1971, 5, 205–208.

Lohr, S., With a whole lot at stake, I.B.M. and Microsoft await year 2000. *NY Times*, C1, December 31, 1999.

Senate Special Committee on the Year 2000 Technology Problem, *Investigating the Impact of the Year 2000 Problem*. February 24, 1999. http://www.senate.gov/~y2k/documents/report/.

Senate Special Committee on the Year 2000 Technology Problem, *Investigating the Year 2000 Problem: The 100 Day Report*. September 22, 1999. http://www.senate.gov/~y2k/documents/100dayrpt/.

Senate Special Committee on the Year 2000 Technology Problem, *Y2K Aftermath—Crisis Averted Final Committee Report*. February 29, 2000. http://www.senate.gov/~y2k/documents/final.pdf.

Sullivan, P., Obituaries: Computer Pioneer Bob Bemer, 84. *Wash Post*, B6, June 25, 2004.

United States Code (USC), 2005. http://www.straylight.law.cornell.edu/uscode.

United States Senate, *U.S. Senate Creates New Committee, Bennett to Serve as Chairman*. April 3, 1998. http://www.senate.gov/~y2k/news/pr980403.htm.

Weingarten, G., The millennium bug. *Wash Post*, F1, July 18, 1999.

Zeller, T., Jr., and Mayersohn, N. Can a virus hitch a ride in your car? *NY Times*, 8–12, March 13, 2005.

QUESTIONS FOR DISCUSSION

1. Should a programmer remain responsible for his code decades after it is coded?

2. When planning a software architecture, what length of time should be considered "short term"? What length of time should be considered "long term"?

3. Based on cost-benefit analysis, when should the first Y2K fixes have been initiated?

4. In the early 2000s, several viruses and worms disrupted daily business at American corporations. In early 2005 the *New York Times* reported that a virus may one day disrupt automobile functions. One mechanism through which this could occur would be to infiltrate the OnStar system, which will be included in all General Motors cars by the end of 2007. OnStar can forward readings from sensors throughout the car through a cell phone link for troubleshooting. Several automakers have discussed plans to use this conduit, called telematics, to update a vehicle's software or even perform electronic repairs (Zeller, 2005). How could a telematics virus affect public safety? Should planning for telematics security become an automotive industry mandate?

5. How involved should the government be in keeping computer systems of all types secure?

Chapter 10

2002: Bell Laboratories Scientific Fraud

THE REPORTED STORY

The *New York Times* Abstract:

Investigation committee finds series of extraordinary advances in physics claimed by scientists at Bell Labs relied on fraudulent data; findings dismiss as fiction results from 17 papers that were promoted as major breakthrough in physics, including claims that Bell Labs had created molecular-scale transistors; committee concludes that data in disputed research, published between 1998 and 2001, was improperly manipulated, even fabricated; this confirms suspicions raised by outside scientists in May; committee places blame for deceit on Bell Lab scientist Dr J Hendrik Schon; Bell Labs immediately fires Schon; just last year he was thought to be on fast path to Nobel Prize; panel finds no other scientists were guilty of misconduct, but scandal has tarnished surrounding participants, including co-authors who noticed nothing amiss, scientific journals that quickly published sensational findings, and Lucent Technologies, Bell Labs' parent company; case raises questions about core of scientific process. (Chang, 2002)

THE BACK STORY

THE HISTORY OF BELL LABORATORIES

In 1876, Alexander Graham Bell invented the telephone; his invention resulted in two issued patents. With two partners, Bell formed a company in 1877 that later became American Telephone and Telegraph (AT&T). When Bell's second patent expired in 1894, competing telephone companies entered the market. By 1904 more than 6000 companies offered telephone service in localities through the United States. However, because networks from these companies were not interconnected, subscribers to different companies could not call each other. To address this lack of interconnection, the U.S. government accepted AT&T's proposal in 1913 that telephone service would be operated most efficiently as a monopoly providing universal service. Under this agreement, called the Kingsbury Commitment, AT&T agreed to connect noncompeting independent telephone companies to its network and to divest its controlling interest in Western Union Telegraph. In 1956, AT&T further agreed to restrict its activities to its national telephone system (both local and long distance service) and government work. On January 1, 1984, as the settlement of an antitrust suit brought by the U.S. government in 1976, AT&T divested itself of the wholly owned Bell operating companies that provided local exchange service. On September 3, 1996, AT&T voluntarily separated out Lucent Technologies as a publicly traded systems and equipment company. Lucent kept the Bell Laboratories name (AT&T, 2004).

Although AT&T officially established Bell Telephone Laboratories as its research and development (R&D) subsidiary in 1925, the roots of telephone research can be traced back to 1885. Managers of the AT&T engineering department formed a research department to investigate the physics of electromagnetic propagation on long distance lines. This new department was headed by Hammond Hayes, one of Harvard's first physics PhDs. In 1899, George Campbell developed the theory of loading coils, which reduced the rate at which a transmitted telephone signal weakened. This practical invention doubled the maximum transmission distance of open lines, allowing the long distance network to extend from New York to Denver by 1911. In 1915, using the first practical electronic amplifier, developed by Harold Arnold, AT&T opened its first transcontinental telephone lines (Lucent, 2004).

From its inception, Bell Telephone Laboratories, which later became known as Bell Laboratories or Bell Labs, was a special place. During its years as a regulated monopoly, the amount of profit AT&T could earn was fixed by law. As AT&T's R&D arm, Bell Labs was treated as part of the cost of maintaining and upgrading the network, with R&D expenditures written off as a business expense. In essence, Bell Labs was a nationally supported laboratory, financed by every phone booth coin and phone bill monthly payment.

Bell Labs hired only the best and brightest and provided an environment for them to excel. Alfred Cho, co-inventor of the molecular beam epitaxy (MBE) machine and a research director in 1991, stated that "in the old days, we went out and hired the best people we could, turned them loose in a large room and said, 'Show us what you can do.' This provided the climate for good science and good technology." His MBE machine, an ultra–high-vacuum crystal growth device able to create one atomic layer at a time and an indispensable piece of equipment for the semiconductor industry, was the result of a 12-year research project (Crease, 1991). This environment produced six Nobel prizes in physics shared by eleven scientists (Lucent, 2004):

1937: wave nature of matter
1956: transistor
1977: improved understanding of local electronic states in solids
1978: radio astronomy (discovery of background radiation remaining
    from the "big bang" explosion)
1997: optical trapping
1998: fractional quantum Hall effect

During the monopoly, this passion for excellence applied to development as well as research. Engineers were recruited from top universities, with the entry-level engineering position requiring a Master's degree and technician position requiring a Bachelor's degree. Products to be developed were specified for optimum performance and high reliability under an extremely wide range of operating conditions without cost constraints. When appropriate, new research technologies were incorporated into proposed products. Products whose specifications were validated through extensive testing were deployed in the telephone network. Robert Lucky, executive director of communications research until 1992 and inventor of adaptive equalization (for correction of telephone channel distortion), recently reminisced about the "golden years" before divestiture in reflections for *IEEE Signal Processing* magazine. He recalled how, during the AT&T antitrust trial, he found himself explaining to Judge Greene "the importance of research at Bell Labs and our guiding religion of trying to create the best telephone network for the country irrespective of economic gain to the company itself" (Lucky, 2004). Bell Labs' engineering accomplishments include the following (Lucent, 2004):

1927: first transatlantic telephone service
1927: first demonstration of television in the United States
1947: invention of cellular telephony
1958: first commercial modem
1962: first active communications satellite
1965: first electronic telephone switch
1971: invention of Unix computer operating system

This unique R&D culture did not change for several years after divestiture in 1984. However, in September 1990 a long-expected reorganization occurred. Research laboratories (initially approximately 125) were reorganized, with some eliminated, some created, and many combined. Basic physics research was scaled back, while software-related research increased. Research was aligned with activities of the business units, with research oriented toward near-term (3 to 5 years) results. Concurrent engineering replaced the old paradigm of sequential development, during which, after investigation, research was transferred to development, which after further investigation was transferred to manufacturing. As vice president of research and Nobel prize winner Arno Penzias explained, basic and applied research were remixed to suit the business end of the company. Bell Labs' existence would depend on AT&T's survival in an intensely competitive marketplace (Crease, 1991).

Bell Labs was forced to further reduce headcount in both research and development in January 1996 as part of an overall AT&T effort to cut 40,000 jobs. In physical science research, 70 of 590 scientists were laid off. When Lucent Technologies split from AT&T later that year, new research director Arun Netravali promised that no further cuts would be made in physical science research. Together, 520 physical scientists and 400 information scientists (with more information scientists to be hired) became the new core research staff of Lucent Bell Labs (Service, 1996).

Lucent was very successful during the telecommunications boom. It became a stock market star by promising revenue growth of 20% a year. Its stock closed at $7.66 on its debut on April 4, 1996 and at $75.00 on December 31, 1999. To continue this growth, it is now known that Lucent made improper deals with customers to meet these sales targets during fiscal year 2000. Because it improperly reported $1.148 billion in revenue and $470 million in pre-tax 2000 income, the Securities and Exchange Commission fined Lucent $25 million (Young, 2004).

When telecom crashed, Lucent "was too bloated with employees and facilities to keep pace" (Berman, 2003). Lucent lost $28 billion over a 24-month period from 2001 to 2003. In Spring 2002, for the first time in its history, Bell Labs managers met to conduct full reviews of its 50 research projects. The projects were ranked according to research quality and relevance to Lucent. Those associated with the lowest ranking projects were laid off. As a result, by mid-2003 the core research staff had decreased to 400 scientists. At this time, Bell Labs president William O'Shea stated that "though smaller than in the past, we continue to do great work across a range of fundamental research disciplines." The *Wall Street Journal* article reporting this quote cited that "one such project would make transistors out of plastic, instead of silicon, greatly reducing costs" (Berman, 2003).

NANOTECHNOLOGY

Nanotechnology refers to the observation and manipulation of materials at the molecular and atomic levels of approximately 1 to 100 nanometers. This atomic-scale fabrication was first proposed by Richard Feynman in 1959 as a "bottom-up" approach, as opposed to the "top-down" approach to which we are accustomed. Because atoms or molecules at surfaces are often reactive, the behavior of nanostructures, which typically have high ratios of surface area to volume, is expected to differ from that of ordinary materials. In the United States, government-sponsored nanotechnology research is conducted under the national nanotechnology initiative (NNI), which was first proposed by President Bill Clinton in 2000. Originally funded by $421 million in 2001 (NNI, 2002), NNI received $849 million in 2004 (OSTP, 2004). Nanotechnology is one of the major research areas at Lucent Bell Labs.

One of the first applications for nanotechnology is the extension of Moore's Law. Moore's Law is based on a 1965 prediction by Gordon Moore. Moore, then director of research and development at Fairchild Semiconductors, predicted an annual doubling of the number of transistors that could be fabricated on a semiconductor chip. Since the late 1970s, this doubling period has occurred every 24 months. However, because it is projected that current technology will reach its limit around 2018, another technology, such as nanotechnology, will need to be implemented to continue this doubling period (Ross, 2003). Some groups are investigating biomolecules for self-assembly of small circuits, because deoxyribonucleic acid (DNA) is a well-defined and predictable construction material (Tseng, 2001). Others have constructed circuits out of individual nanotubes or nanowires (tube or wire structures created through self-assembly of atoms) (Tseng, 2001). A third approach, taken by groups such as that at Bell Labs, is to build circuits atom by atom.

JAN HENDRIK SCHON'S BELL LABS WORK

Jan Hendrik Schon joined Bell Labs in 1998 at the invitation of Bertram Batlogg. Working with Batlogg and Christian Kloc, Schon began to study electrical charge conduction through organic crystals. In a series of groundbreaking journal articles starting in February 2000, they demonstrated that they could use devices called field effect transistors (FETs) to inject large numbers of electrical charges into organic (carbon-containing) materials. An FET is a device that consists of a source and drain, which are two electrode-conducting regions, in isolation from each other and from a third region called the gate. The source and drain are semiconductors—that is, materials with intermediate electrical conductivity between a metal, which is a true conductor, and insulator, which is a nonconductor. Because the source and drain are deposited above another type of semiconductor, the flow of electrons from the source to drain may be controlled by a voltage applied to the gate (Figure 10.1a).

Each of Schon's FETs was constructed by growing a millimeter-sized single organic crystal as a substrate and depositing two metal electrodes (source and drain) on top. These two materials were only held together by weak molecular van der Waals forces. A thin insulating barrier of aluminum oxide was deposited to isolate the electrodes above the crystal. A gate electrode was then deposited above the aluminum oxide (Figure 10.1b). Amazingly, when the electrodes were connected to a power supply, the fragile insulator changed to a semiconductor, conducting current when prompted by the gate voltage (Service, 2002b).

By changing charge concentration, Schon tuned the electronic properties of the materials to behave as insulators, semiconductors, metals, or superconductors (conductors without resistance to electrical current). The group also reported that organic FETs displayed superconductivity at a temperature higher than had ever been seen in an organic material, revealed quantum signatures never before seen in organics, and could be made to act as lasers (light amplification by stimulated emission of radiation) and novel superconducting switches (Service, 2002a).

These astounding results were even more remarkable because they were achieved at an incredible rate. Typically, publishing two or three articles a year is productive. But Schon was lead author on dozens of articles, and published 100 articles in 3 years. Most of these articles appeared in leading journals, including *Science* and *Nature*. In 2001, Schon received an award for scientific "breakthrough of the year." Less than 5 years after finishing graduate school, Schon was in contention for the Nobel prize in physics (Cassuto, 2002).

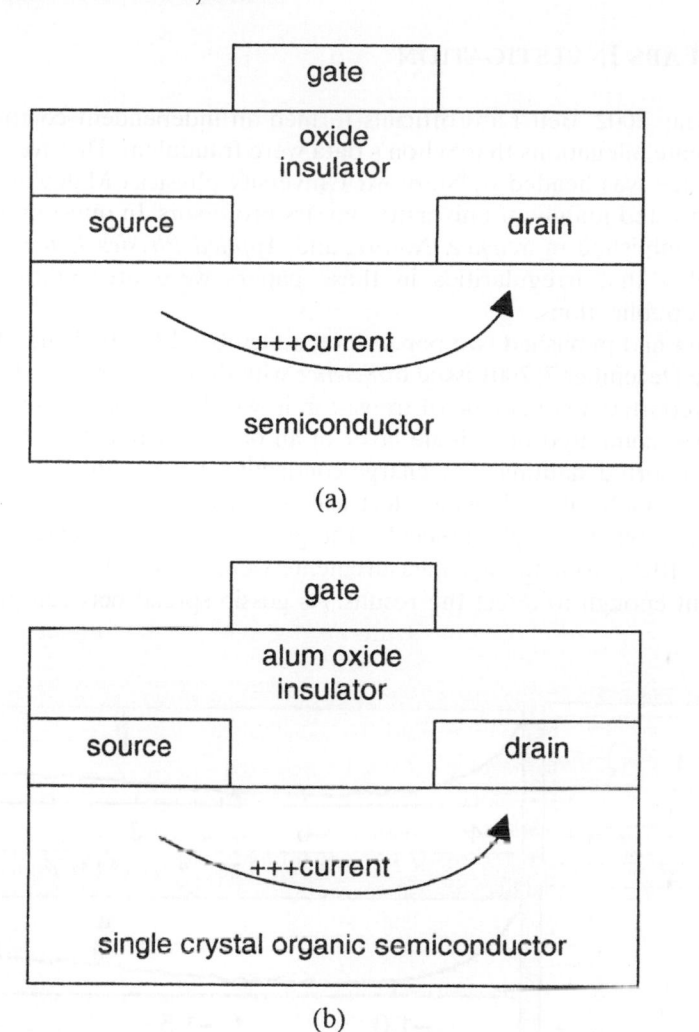

Figure 10.1   (a) Semiconductor and (b) organic field effect transistors.

An odd sidebar during this flurry of activity was that no other scientist could duplicate Schon's work. Schon claimed his devices could contain large voltage differences, several times larger than were seen in other scientists' devices. These larger voltage differences were apparently responsible for the new effects. It was estimated that 100 laboratories in the United States and around the world were working on Schon's results by 2002 but could not duplicate them. Tens of millions of dollars, including funding from the U.S. Department of Energy, were spent on this effort (Cassuto, 2002).

BELL LABS INVESTIGATION

In May 2002, Bell Labs officials formed an independent committee to investigate allegations that Schon's data were fraudulent. The five-member committee was headed by Stanford University physicist Malcolm Beasley and consisted mainly of university physics professors. In question were six papers published in *Science, Nature*, and *Applied Physics Letters* and the possibility that irregularities in these papers were present throughout Schon's publications.

Schon had published two papers in the October 18, 2001 issue of *Nature* and the December 7, 2001 issue of *Science* with duplicate figures. The *Nature* paper reported a novel type of transistor, in which a key charge-conducting layer was composed of a single layer of an organic conductor. The *Science* paper reported diluting that charge-conducting layer with nonconducting insulating molecules, allowing electrical conductivity in a transistor to be tracked through a single molecule. The graphed results appeared identical (Figure 10.2), even though measurements were recorded at temperatures different enough to affect the results. As gossip spread between physicists,

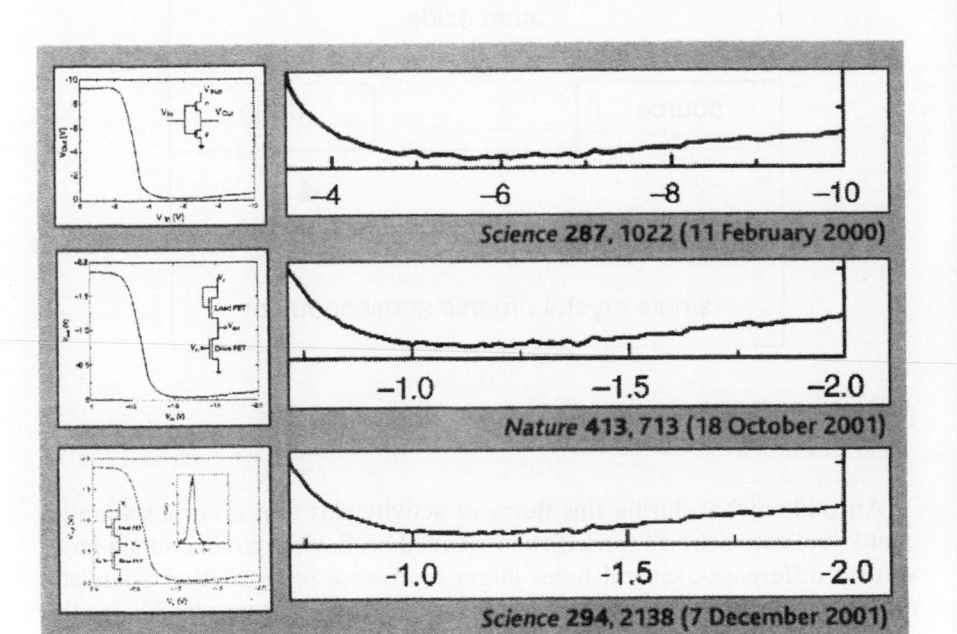

Figure 10.2 Identical noise levels in figures from three Schon papers.
Reprinted with permission from Service, 2002a. Copyright 2002 AAAS. Reprint permission also obtained from the Nature Publishing Group.

Schon e-mailed *Science* that an error had occurred, asking that a new figure be printed to correct the wrong figure mistakenly incorporated into the *Science* paper.

However, the noise levels in these figures, which should have appeared random compared with each other, also appeared similar to the noise levels in a third figure published in a paper in the February 11, 2000 issue of *Science* (see Figure 10.2). The third paper reported a device with a different material in the key charge-conducting channel of the FET and a different geometry. Both these differences should have caused this device to behave differently over the operating (non-noise) range from the devices in the two 2001 papers, not similarly as reported.

Further, two other figures in the February 11, 2000 *Science* and October 18, 2001 *Nature* papers appeared similar to each other and to a figure that appeared in the April 28, 2000 issue of *Science*. Again, these three papers described different organic conductors, and no physical reason exists why they would be so similar. Two other papers, appearing in the November 3, 2000 issue of *Science* and the December 4, 2000 issue of *Applied Physics Letters*, also had identical figures that illustrated unexpected physical results (Service, 2002a).

Over the course of three months, the committee discovered that the inconsistencies in these six papers reflected much larger problems of data substitution in nine papers, unrealistic precision in nine papers, and contradictory physics in six papers. *Data substitution* referred to substitution of whole figures, single curves, and partial curves in different or the same paper to represent different materials, devices, or conditions. *Unrealistic precision* referred to precision beyond that expected in a real experiment or requiring unreasonable statistical probability. *Contradictory physics* referred to behavior inconsistent with stated device parameters and prevailing physical understanding, so as to suggest possible misrepresentation of data. Complicating the committee's analysis was the lack of physical evidence. Schon had not maintained any systematic records such as a signed notebook, and had deleted all primary electronic data files because of "insufficient storage capacity." He had no working device to confirm claimed results, as they had either been damaged during measurement, damaged in transit between the labs in New Jersey and Germany where he conducted his work, or simply discarded. Further, with one exception reportedly witnessed by Bertram Batlogg, no measurement or demonstration of a significant physical effect or device characteristic was witnessed by any co-author or colleague (Lucent, 2002).

The committee released their findings on September 26, 2002. Of the 24 allegations, the committee concluded conclusively that Schon was guilty of scientific misconduct on 16 counts. Schon was fired later that day and was forced to return immediately to Germany because his U.S. visa was contingent upon U.S. employment. The committee also cleared all co-authors of any scientific misconduct. However, they questioned the professional responsibility of Schon's supervisor, Bertram Batlogg, in not validating Schon's results before publication (Lucent, 2002).

AFTERMATH

After the committee released its findings, many began to question why Schon's work had not received further internal scrutiny before publication. According to Bell Labs spokesman Saswato Das, "In our view, this was an isolated, anomalous incident." But John Rowell, former director of chemical physics at Bell Labs who worked there from 1961 to 1983, disagreed. Stated Rowell, "This is certainly not the way things used to be at Bell Labs. In the good old days, experiments would be immediately witnessed by one or sometimes even two levels of management, and by collaborators. . . . There was a rigorous publication release process that involved circulation of papers to management and other researchers" (Lerner, 2002).

The peer-review process was also questioned. But according to Monica Bradford, managing editor of *Science*, "After the story broke, we looked back over the reviewer reports, but we did not find any clues that something was wrong." Said Karl Ziemelis, physical sciences editor at *Nature*, "Clearly, reviewers were less critical of the papers than they should have been, in part because the papers came from Batlogg, who had an excellent track record, and from Bell Labs, which has always done good work. . . . In addition, although the results were spectacular they were in keeping with the expectations of the community. If they had not been, or had they come from a completely unknown research group, they might have gotten closer scrutiny" (Lerner, 2002). As part of its manuscript submission process, *Science* now asks that if a paper is accepted for publication, then "any reasonable request for materials, methods, or data necessary to verify the conclusions of the experiments reported must be honored" and that large data sets be deposited in an approved database or housed as supporting online material at *Science* (AAAS, 2004).

APPLICABLE REGULATIONS

The Bell Labs investigation committee used the U.S. Federal Policy on Research Misconduct (OSTP, 2003) as its guiding set of principles, definitions, and recommended practices in conducting its investigation.

Although the research in question was not federally funded (and thus this policy did not affect Bell Labs), this policy represents a consensus view of the U.S. scientific community on the issue of scientific misconduct (Lucent, 2002).

This policy contains six parts: research conduct definitions, findings of research misconduct, responsibilities of federal agencies and research institutions, guidelines for fair and timely procedures, agency administrative actions, and roles of other organizations. According to Parts I and II, research misconduct and findings of research misconduct are defined as the following:

I. Research Misconduct Defined

Research misconduct is defined as fabrication, falsification, or plagiarism in proposing, performing, or reviewing research, or in reporting research results.

- *Fabrication* is making up data or results and recording or reporting them.
- *Falsification* is manipulating research materials, equipment, or processes, or changing or omitting data or results such that the research is not accurately represented in the research record.
- *Plagiarism* is the appropriation of another person's ideas, processes, results, or words without giving appropriate credit.
- Research misconduct does not include honest error or differences of opinion. (OSTP, 2003)

II. Findings of Research Misconduct

A finding of research misconduct requires that:

- There be a significant departure from accepted practices of the relevant research community; and
- The misconduct be committed intentionally, or knowingly, or recklessly; and
- The allegation be proven by a preponderance of evidence (OSTP, 2003). Per this policy, it is the responsibility of the home research institution to conduct an investigation into allegations of scientific misconduct. If misconduct is found, then the federal agency that funded this research shall determine the seriousness of the misconduct and the appropriate administrative actions. An administrative action may range from a letter of reprimand to a suspension or termination of an active funding award. If the funding agency believes criminal or civil fraud violations may have occurred, it shall refer the matter to the Department of Justice, the Inspector General for the agency, or other appropriate investigative body. (OSTP, 2003)

A SCIENTIFIC PERSPECTIVE

As Schon's molecular electronics results were published, colleagues began questioning his results. According to physicist Leo Kouwenhoven, who worked at Bell Labs during this period, "People were always gossiping about Hendrik" (Delta, 2002). Said Princeton physicist Lydia Sohn, "Collectively, people at Bell [Labs] were nervous" (Service, 2002a).

Although Schon's initial papers did not provide much experimental detail, it was expected that more details would be provided with future results. However, as Dutch physicist Teun Klapwijk later admitted, physicists became irritated "that Schon's flurry of papers continued without increased detail, and with the same sloppiness and inconsistencies" (Cassuto, 2002). Kouwenhoven's colleagues formed an e-mail group, exchanging information on inconsistencies in Schon's work (Delta, 2002). Several anonymous Bell Labs researchers were the first to discover the discrepancies in the *Science* and *Nature* papers and relayed this information to Sohn in April (Service, 2002a). Sohn and Cornell physicist Paul McEuen then conducted an initial analysis of all Schon's papers, eventually discovering the data similarities in six papers. Said Sohn, "The data were too clean. They were what you'd expect theoretically, not experimentally. People were getting frustrated because no one could reproduce the results, and it was hitting a crescendo" (Cassuto, 2002).

On May 10, 2002, McEuen and Sohn contacted Schon, Batlogg, managers at Bell Labs, and manuscript editors at *Science* and *Nature*, disclosing the data similarities in three papers (similarities in the other three papers were discovered a few days later). An independent committee was then appointed by Bell Labs to investigate Schon's work (Service, 2002a).

Lydia Sohn never disclosed the names of the anonymous Bell Labs researchers who contacted her about discrepancies in Schon's papers (Service, 2002a).

REFERENCES

American Academy for Advancement of Science, *Science's Electronic Manuscript Submission Website*. Washingon, DC: AAAS, 2004. http://www.submit2science.org/ws/menu.asp.

AT&T, *The History of AT&T*. Basking Ridge, NJ: AT&T, 2004. http://www.att.com/history.

Berman, D., New calling: at Bell Labs, hard times take toll on pure science. *WSJ*, A1, May 23, 2003.

Cassuto, L., Big trouble in the world of "big physics." *Salon*, Sept. 16, 2002. http://www.salon.com.

Chang, K., Panel says Bell Labs scientist faked discoveries in physics. *NY Times*, A1, C2, September 26, 2002.

Crease, R., Bell Labs: shakeout follows breakup. *Science*, 1991, 252, 1480–1482.

Delta, "Finally, we've caught Hendrik." *Delta* (online newsletter for Technical University of Delft), August 29, 2002, 34, 4.

Lerner, E., Fraud shows peer-review flaws. *The Industrial Scientist*, 2002, 8, 12. http://www.aip.org/tip/INPHFA/vol-8/iss-6/p12.html.

Lucent Technologies, *The Exciting World of Bell Labs*. Murray Hill, NJ: Lucent Technologies, 2004. http://www.bell-labs.com/history.

Lucent Technologies, *Report on the investigation committee on the possibility of scientific misconduct in the work of Hendrik Schon and coauthors*. Murray Hill, NJ: Lucent Technologies, 2002. http://www.lucent.com/news_events/pdf/researchreview.pdf.

Lucky, R., Leadership and life in the old Bell Labs. *IEEE Sig Proc Mag*, May, 2004, 21, 6–8.

National Nanotechnology Initiative, *Research and Development FY 2002*. Washington, DC: NNI, 2002. http://www.nano.gov/2002budget.html.

Office of Science and Technology Policy, *U.S. Federal Policy on Research Misconduct*. Washington, DC: OSTP, 2003. http://www.ostp.gov/html/001207_3.html.

Office of Science and Technology Policy, *National Nanotechnology Initiative*. Washington, DC: OSTP, 2004. http://www.ostp.gov/html/budget/2004/OSTP%20NNI%201-pager%20(OMB).pdf.

Ross, P., 5 commandments. *IEEE Spectrum*, 2003, 40, 30–35.

Service, R., Relaunching Bell Labs. *Science*, 1996, 272, 638–639.

Service, R., Pioneering physics papers under suspicion for data manipulation. *Science*, 2002a, 296, 1376–1377.

Service, R., Winning streak brought awe, and then doubt. *Science*, 2002b, 297, 34–37.

Tseng, G., and Ellenbogen, J., Towards nanocomputers. *Science*, 2001, 294, 1293–1294.

Young, S., and Berman, D., Lucent settlement unveiled by SEC: 10 face civil suits. *WSJ*, A3, May 18, 2004.

QUESTIONS FOR DISCUSSION

1. Is a supervisor responsible for his direct report's work?

2. What is the threshold for naming someone who works on a project as a co-author? Is a co-author responsible for the main author's work?

3. If Schon's fraudulent papers had all been submitted for publication after 2003, could he be prosecuted under the new Sarbanes-Oxley Act (see Chapter 2)?

4. Read the following opinion article describing the peer review process, which was written by veteran researcher Howard Birnbaum: http://www.physicstoday.org/vol-55/iss-3/p49.html. It is titled "A Personal Reflection on University Research Funding." How can the fairness of peer review funding be ensured?

5. Manuscript reviewers may be biased toward prestigious groups and accepted ideas. One suggested reform to minimize bias is blind review; that is, removing the authors' names from articles sent to reviewers. Another suggested reform is open review; that is, reviewer identification in reviews seen by the authors. What are the pros and cons of blind review and open review?

Chapter 11

2002: Ford Explorer Rollover

THE REPORTED STORY

The *San Diego Union Tribune* Abstract:

> Ford Motor Co. will pay state attorneys general $51 million to end claims
> that its advertising fails to disclose the rollover risk involved with driving sport
> utility vehicles, the Associated Press has learned. The money will be shared
> among the 50 states, the District of Columbia, Puerto Rico and the Virgin
> Islands, said four sources who spoke on condition of anonymity. (AP, 2002)

THE BACK STORY

FORD EXPLORERS AND FIRESTONE TIRES

In May 2001, Ford Motor Company recalled and replaced 13 million
Firestone tires equipped in its Ford Explorer sport utility vehicles (SUVs).
This action was an attempt to end its dispute with Bridgestone/Firestone Inc.
over who was to blame for deadly Explorer rollover crashes. The dispute had
already generated bad publicity, launched congressional hearings, and
spawned a new focus on tire safety. In part because of the $3 billion price tag
for replacing the tires, Ford lost $5.45 billion in 2001 (Webster, 2003).

However, Firestone tires seem unlikely to have been the fundamental
cause of Ford Explorer rollover crashes. During the 10-year period during
which Ford–Firestone-related rollovers caused approximately 300 deaths,

more than 12,000 people died in SUV rollover crashes (all models, not just the Explorer) unrelated to tire failure (Frontline, 2002a). In this chapter, we discuss issues connected to the design of the Ford Explorer SUV.

CAFE STANDARDS AND SPORT UTILITY VEHICLES

In response to the 1973 Arab oil embargo, which increased gasoline prices and fuel shortages, the U.S. Congress rushed to pass fuel efficiency legislation for cars and light trucks. This legislation was included in the Energy Policy and Conservation Act, which passed in 1975. Per the act, automakers were required to meet strict fuel efficiency standards known as corporate average fuel economy (CAFE). CAFE standards were administered by the National Highway Traffic Safety Administration (NHTSA). The near-term goal was to double new passenger car fuel efficiency to 27.5 miles per gallon (mpg) by model year 1985. Because light trucks were primarily used on farms and ranches, light truck fuel efficiency was to be set at the maximum feasible level for model year 1979 and each model year thereafter. Failure to meet the CAFE standard would result in fines, based on the total number of vehicles produced by an automaker in a given model year (NHTSA, 2005a).

With CAFE standards in place, fuel economy rose and helped turn an oil shortage into a glut (Kennedy, 2004). CAFE standards were decreased from model years 1986 to 1989 and fixed thereafter to 27.5 mpg (NHTSA, 2005a). Had CAFE standards continued at their original rate of increase through the end of calendar year 1986, it has been estimated by economist Amory Lovins of the Rocky Mountain Institute that the United States would not have needed Persian Gulf oil after 1986 (Kennedy, 2004).

Automakers took advantage of the light truck "loophole" to manufacture passenger cars that did not have to comply with CAFE standards. In 1975, American Motors updated its Jeep CJ5 with modern 1970s styling. The original Jeep was a military truck built for World War II usage. As the Jeep CJ5 became popular and gas prices decreased in the early 1980s, American Motors and other manufacturers debuted more sports utility vehicles. New SUVs such as the Jeep Cherokee, Ford Bronco II, and Ford Explorer provided large profits to ailing 1980s Detroit automakers because they were essentially pickup truck parts sold for a luxury car price. Different passenger compartments were bolted onto the steel underbodies of pickup trucks, enabling pickups and SUVs to be built on the same assembly line (Frontline, 2002d).

THE FORD BRONCO II

When Ford market released the Bronco II in March 1983, it was marketed as a vehicle for men, single people, or young couples. As described by Ford marketer Martin Goldfarb, "I think sport utility vehicles were almost like John Wayne vehicles. It was the excitement of discovery, the excitement of America, the rugged individual. . . . The Bronco II was this four-wheel drive vehicle that gave people the sense they could conquer anything; it could go anywhere" (Frontline, 2002e).

To minimize time to market, Ford chose to modify its existing Ford Ranger platform rather than to design from scratch. A new design would have delayed production for 1 to 2 years, placing Ford more than 1 year behind the release of General Motor's competitive SUV, the Chevrolet S-10 Blazer. The use of the Ranger platform also saved $300 million, because both could be manufactured on the assembly line. The profit for each Bronco II was estimated to be $3750 (Opinion, 1999). Eventually, 700,000 Bronco IIs were produced (Frontline, 2002b).

But during the Bronco II's 1981 design phase and 1982 verification phase, stability problems began to surface. Eight months before production began, Ford's Office of the General Counsel, for the first time in Ford's history, collected all documents related to a product's handling characteristics. One hundred thirteen documents were collected that were specifically related to the Bronco II program reports, test requests, test plans, and simulation analysis. Fifty-three of these documents disappeared (Opinion, 1999). This unusual document handling procedure foreshadowed the Bronco II lawsuits filed against Ford beginning in the late 1980s.

THE FORD EXPLORER

According to Goldfarb, the Ford Explorer "was a vehicle to replace the Bronco II, or to grow from the Bronco II, because the Bronco II had a limited audience. . . . Well, this was going to be a family vehicle, but gave you that same sport environment, that same outdoor feeling that a Bronco II did. . . . Cars like Explorer exemplify the desire to do physical things that had a dangerous tonality to it, like skiing, and Explorer was part of that" (Frontline, 2002e). Months after it went on sale in April 1990, it was America's number one sport utility vehicle (Figure 11.1). By the end of 2000, there were 3.2 million Explorers on the road. The profit on each explorer was nearly $8000.

Figure 11.1   2000 Ford Explorer after National Highway Safety and Traffic Administration (NHTSA) side crash testing.
Courtesy NHTSA, http://www.safecar.gov.

Unfortunately, the Explorer design was also based on the Ranger underbody. During its early design phase, which started in 1986, Explorer developers had no knowledge of Bronco II stability problems. In early 1989, a year away from mass production, *Consumer Reports* criticized the Bronco II's tendency to roll over during certain high-speed turns. Explorer engineers then proposed three design options to improve stability:

1. Shorter suspension springs to lower the front end by half an inch and the back by an inch.
2. Fairly low tire pressure, which would give the Explorer a more stable ride (except when a tire failed), as well as the softer ride favored by people accustomed to cars.
3. Entire vehicle redesign to mount the wheels 2 inches farther apart.

As the Bronco II managers had done, Explorer managers chose options that would not delay scheduled market release. Options one and two were implemented. But all the Explorer design decisions came with tradeoffs. By extending the passenger compartment and installing a second row of seats, designers made the Explorer 600 pounds heavier than the Ranger, without upgrading the suspension and tires to carry the bigger load. Overloading decreases stability and handling. By choosing the same size tires used in the Ranger, designers inherited tires that had the lowest possible rating for withstanding high temperatures. By lowering the recommended tire pressure to increase stability and soften the ride, the tire's ability to carry weight without overheating was further reduced. Because all pickup-based designs rely on two stiff, heavy steel beams that run the length of the underbody and curve up like runners of a sleigh, these beams tend to slide up and over cars' bumpers and door sills, punching into the other vehicle's passenger compartment (Bradsher, 2000).

GOVERNMENT REGULATION OF SUVS

The National Highway Traffic Safety Administration was very active from the time of its creation in the 1960s through the Carter administration. During this period, it crafted new standards and investigated the General Motors Corvair and the Ford Pinto. Unfortunately, under Ronald Reagan, the power of NHTSA was sharply diminished by severe budget and staff cuts. Its funding in 2001 remained less than one-third of its 1980 funding, after adjusting for inflation (Frontline, 2002c).

In 1986, Congressman Tim Wirth, who was chairman of the House subcommittee overseeing NHTSA, petitioned NHTSA to craft a rollover prevention standard and to conduct a rollover defect investigation of certain vehicles. According to former NTHSA administrator Joan Claybrook, who served under Jimmy Carter, NHTSA staff wanted to comply with the Wirth petition (Frontline, 2002c). NHTSA engineer Anna Harwin, who was assigned to the initial investigation, even found a consistent relationship between center-of-gravity height and track width versus rollover (Frontline, 2002b). The calculation of this static stability factor is shown in Figure 11.2.

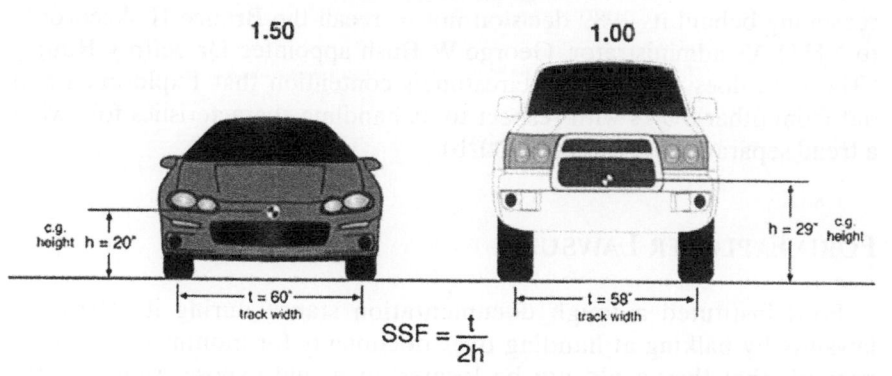

Rollover Ratings
Static Stability Factor

$$SSF = \frac{t}{2h}$$

Figure 11.2 Static stability factor calculation. On average, these are how vehicles are ranked in increasing order of static stability factor and decreasing order of rollover probability: sport utility vehicle, pickup truck, van, car.
Reprinted from NHTSA, 2005b.

A related parameter, the stability index, is calculated as track width divided by height. Unfortunately, Claybrook's successor, Reagan appointee Diane Steed, declined to act and turned down the petition (Frontline, 2002c).

Eventually, with more models of SUVs with rollover propensities on the road, NHTSA was forced to investigate rollover defects. Specifically, it conducted a defect investigation of the Bronco II from late 1988 to October 1990. However, NHTSA's investigation was incomplete because Ford held back key Bronco II stability testing failure documents that were requested by NHTSA. NHTSA neither conducted tests on its own nor questioned Ford about the documents produced. In its closing report, NHTSA stated that there "appears to be no reasonable expectation that further investigation would lead to a determination of the existence of a safety-related defect with respect to any of the allegations regarding the propensity of the Bronco II to rollover" (Opinion, 1999). According to Claybrook, "When the agency refused to do a recall of that vehicle, it gave a pass to every other SUV" (Frontline, 2002c).

The Ford-Firestone crisis enabled NHTSA to regain some of its authority. In reaction to the Ford-Firestone crisis, Congress passed the Transportation Reporting Enhancement, Accountability and Documentation (TREAD) Act of 2000. Besides requiring auto and tire manufacturers to report defects on American autos and tires sold in foreign countries, TREAD increased NHTSA's authority to collect information about possibly defective products and expanded its budget for investigations. Further, by the end of 2002 NHTSA was required to produce a dynamic stability test for rollovers.

In a related decision seen as a victory for Ford, NHTSA denied Bridgestone/Firestone's request to investigate safety defects in the Ford Explorer. NHTSA's reasoning behind this decision was the same as the reasoning behind its 1989 decision not to recall the Bronco II. According to NHTSA's administrator, George W. Bush appointee Dr. Jeffrey Runge, "The data does not support Firestone's contention that Explorers stand out from other SUVs with respect to its handling characteristics following a tread separation" (Frontline, 2002b).

FORD EXPLORER LAWSUITS

Ford instituted a tough documentation stance during its Explorer lawsuits by balking at handing over documents for months or years on grounds that they could not be located or would expose trade secrets. Typically, during automotive lawsuits, plaintiffs try to sift through documents for evidence that manufacturers may have cut corners to shave costs, or ignored safety recommendation of engineers. According to Frank Branson, a Texas lawyer who has often tangled with Ford, he "can't remember encountering a defendant who set about in a more orchestrated way to conceal evidence from the public's eye and from disclosure in courtrooms." Additionally, judges who have heard Ford cases describe Ford's courtroom conduct as "totally reprehensible," "disgusting," "blatantly lied," and "almost borders on the criminal." Tom Feahency, a Ford vice president in the 1970s and 1980s who now testifies as an expert witness against Ford, agrees. "They've been an outstanding practitioner of it [so that] an awful lot of plaintiffs go away" (Levin, 2004a).

With this litigation strategy, Ford literally settled hundreds of cases involving Explorer rollovers, and won 13 trials in which plaintiffs claimed Explorers were defective because of rollover risk or inadequate roof risk. Recall that Explorers with low static stability factors had been manufactured from 1990 until 2000. Not until June 2004 did Ford lose its first Explorer trial, in which a jury found an Explorer was defective because of its instability and weak roof. A jury awarded Benetta Buell-Wilson $246 million in punitive damages and $122.6 million in compensatory damages, for a total of $368.6 million. In January 2002, Buell-Wilson was paralyzed after she lost control of her 1997 Explorer (Levin, 2004b). Ford then lost three subsequent Explorer rollover trials in August 2004; on March 1, 2005; and on March 18, 2005. Numerous other lawsuits are pending. As an indication of total lawsuit size, Ford lawyers acknowledged there were more than 1600 lawsuits or claims involving Explorer rollovers before July 2000 (Plungis, 2005).

THE 2002 FORD EXPLORER

Between 1983 and 2001, 3,826 people were killed in Ford Bronco II or Explorer rollovers (EWG, 2005a). In consideration of safety, the 2002 model Ford Explorer was completely redesigned. Its tires were larger, with a higher recommended pressure, and it was 2.5 inches wider. The primitive leaf-spring suspension dating back to buggies was replaced by carlike coil springs, which improved braking and were more resistant to sideways wheel movements. Frame rails were now enclosed, rather than shaped like the letter "C," which made the vehicle stiffer in general (Bradsher, 2000).

APPLICABLE REGULATIONS

California was one of plaintiffs in the $51.5 settlement related to Ford Explorer false advertising claims, which were described at the beginning of this chapter. The False Advertising statute for California is given here:

BUSINESS & PROFESSIONS CODE

Division 7 General Business Regulations

§ 17500. False or misleading statements generally

It is unlawful for any person, firm, corporation or association, or any employee thereof with intent directly or indirectly to dispose of real or personal property or to perform services, professional or otherwise, or anything of any nature whatsoever or to induce the public to enter into any obligation relating thereto, to make or disseminate or cause to be made or disseminated before the public in this state, or to make or disseminate or cause to be made or disseminated from this state before the public in any state, in any newspaper or other publication, or any advertising device, or by public outcry or proclamation, or in any other manner or means whatever, including over the Internet, any statement, concerning that real or personal property or those services, professional or otherwise, or concerning any circumstance or matter of fact connected with the proposed performance or disposition thereof, which is untrue or misleading, and which is known, or which by the exercise of reasonable care should be known, to be untrue or misleading, or for any person, firm, or corporation to so make or disseminate or cause to be so made or disseminated any such statement as part of a plan or scheme with the intent not to sell that personal property or those services, professional or otherwise, so advertised at the price stated therein, or as so advertised. Any violation of the provisions of this section is a misdemeanor punishable by imprisonment in the county jail not exceeding six months, or by a fine not exceeding two thousand five hundred dollars ($2,500), or by both that imprisonment and fine. (California, 2005)

AN ENGINEERING PERSPECTIVE

As already discussed earlier, stability problems began to surface during the Bronco II's 1981 design phase and 1982 verification phase. The design goals to "reduce rollover propensity" and "respond safely to large steering inputs which are typical of accident avoidance or emergency maneuvers" could not be met during this project. In 1981, Ford engineers suggested five proposals to increase Bronco II stability. Adopting proposals three, four, or five would have caused the schedule to slip past the March 1983 release. Adopting the fifth proposal would additionally have resulted in $13.8 million in retooling costs and $54 per vehicle in piece costs. Instead, Ford chose to proceed with the second proposal, understanding that this would only increase the stability index to 2.03, worse than the Jeep CP7 stability index of 2.04. The Jeep CP7, related to the Jeep CP5, was Ford's chosen SUV for emulation.

After mechanical prototypes were built, verification testing began. Performing the standard J-turn test (keeping the steering wheel straight, then turning it and holding it in the turn) caused rollovers at speeds as low as 30 miles per hour (mph). Engineers tried various combinations of suspensions, tires, and steering designs to increase stability, but they failed. By mid-March 1982, they reported that the track width needed to be increased by 3 to 4 inches to improve stability. This recommendation, even when revised to be a shorter width increase of 2 inches, was rejected because it would delay market release by 3 months. Eight months before market release, J-turn maneuver testing of prototype Bronco IIs at the Arizona Proving Ground was abandoned after test drivers experienced prototypes tipping up onto two wheels, outriggers failing, and vehicles pole-vaulting. Managers became concerned over the safety of their test drivers. Engineers again recommended increasing the track width by 2 inches, which would delay market release, but were denied. Instead, Ford implemented superficial changes to improve stability, including adding weight beneath the center of gravity, adding sealant to the tires, and changing the wheels.

In September 1986, because engineers continued to raise safety concerns, Ford considered using larger tires. However, it disregarded this idea as it would decrease the stability index and raise questions with the Ford Office of General Counsel (Opinion, 1999).

This section would not be complete without disclosing that a key Ford engineer received $5 million from Ford in return for false court testimonies. Of all 13 cases discussed in this textbook, this is the only case, found by the author, of a staff engineer or technician blatantly disregarding his professional responsibilities of protection of public safety, technical competence, and timely communication of negative and positive results to management.

As the Ford Office of General Counsel reviewed Bronco II documents in anticipation of lawsuits, it noticed that the name "David Bickerstaff" kept appearing. Bickerstaff joined Ford in the early 1970s as a light trucks engineer. He authored a series of engineering reports and memoranda that detailed Ford engineers' concerns that the Bronco II's low stability index would result in rollover accidents. Because Ford suspected Bickerstaff would be called as an expert witness, it offered Bickerstaff and his consulting firm $4000 per day to testify in Ford's favor in 1990. Bickerstaff became Ford's "star" witness in its Bronco II lawsuits. Consequently, Bickerstaff's fees increased over 8 years of court cases, eventually amounting to $5 million. After Ford and Bickerstaff were convicted of engaging in conspiracy to commit fraud at one of the Bronco II trials in 2001, Bickerstaff fled to his native England (EWG, 2005b).

REFERENCES

Associated Press, Ford Motor Co. will pay state attorneys general $51 million to end claims that its advertising fails to disclose the rollover. *SD UT*, A6, December 20, 2002.
Bradsher, K., Risky decision: a special report; study of Ford Explorer's design reveals a series of compromises. *NY Times*, A1, December 7, 2000.
California, *California False Advertising Statute*, 2005. http://www.arb.ca.gov/bluebook/bb04/bus17500/bus_17500.htm.
Environmental Working Group, *Fatal Ford Bronco II/Explorer Rollover Statistics 1983–2001*. Washington, DC: EWG, 2005a. http://www.ewg.org/reports/upsidedown/b2expfatals.php.
Environmental Working Group, *Ford's Fraud on the Court: David Bickerstaff*. Washington, DC: EWG, 2005b. http://www.ewg.org/reports/upsidedown/index5.php.
Frontline, Before you buy an SUV, in *Rollover: the Hidden History of the SUV*, February, 2002a. http://www.pbs.org/wgbh/pages/frontline/shows/rollover/etc/before.html.
Frontline, Chronology: a regulatory free ride, in *Rollover: the Hidden History of the SUV*, February, 2002b. http://www.pbs.org/wgbh/pages/frontline/shows/rollover/unsafe/cron.html#tread.
Frontline, Joan Claybrook interview, in *Rollover: the Hidden History of the SUV*, February, 2002c. http://www.pbs.org/wgbh/pages/frontline/shows/rollover/interviews/claybrook.html.
Frontline, Keith Bradsher interview, in *Rollover: the Hidden History of the SUV*, February, 2002d. http://www.pbs.org/wgbh/pages/frontline/shows/rollover/interviews/bradsher.html.
Frontline, Martin Goldfarb Interview, in *Rollover: the Hidden History of the SUV*, February, 2002e. http://www.pbs.org/wgbh/pages/frontline/shows/rollover/interviews/goldfarb.html.
Levin, M., Ford stonewalls on evidence, judges say. *LA Times*, C1, March 28, 2004a.
Levin, M., Jury adds punitive award in Ford case. *LA Times*, C1, June 4, 2004b.
Kennedy, R. F., Jr., *Crimes Against Nature: How George W. Bush and His Corporate Pals are Plundering the Country and Hijacking Our Democracy*. New York: HarperCollins, 2004, 108. http://www.ewg.org/reports_content/upsidedown/pdf/02581.pdf.
National Highway Traffic Safety Administration, *CAFE Overview—Frequently Used Questions*. Washington, DC: NHTSA, 2005a. http://www.nhtsa.dot.gov/cars/rules/cafe/overview.htm.
National Highway Traffic Safety Administration, *Static Stability Factor Calculation*. Washington, DC: NHTSA, 2005b. http://www.safercar.gov/Rollover/pages/faqs.htm#howisa.
Opinion, *Ford Motor Company v. Vicki Ammerman*, Case No. 49A05-9608-CV-322, Court of Appeals of Indiana, Fifth District, 1999. http://www.ewg.org/reports_content/upsidedown/pdf/AmmermanHL.pdf.
Plungis, J., Memos: Ford made Explorer roof weaker—automaker says SUV exceeds federal safety standards and is a safe vehicle. *Detroit News*, 1A, March 29, 2005.
Surowiecki, J., In case of emergency, *New Yorker*, June 13 & 20, 2005, 70.
United States Code (USC), 2005. http://straylight.law.cornell.edu/uscode.
Webster, S. A., Overcoming difficulties is part of Ford legacy—Explorer rollovers, Pinto explosions are part of journey. *Detroit News*, 7F, June 9, 2003.

QUESTIONS FOR DISCUSSION

1. Describe the similarities among the Ford Pinto, Bronco II, and Explorer design decision-making processes. Why did Ford not seem to learn from its Pinto experience?

2. Rollover problems have plagued sport utility vehicles since primitive truck-based ones were built for the military during World War II. Indeed, costly rollover lawsuits involving truck-based Jeeps pushed American Motors to design the Jeep Cherokee from scratch as a sport utility vehicle in the early 1980s. Based on *New York Times* analysis of federal crash statistics from 1991 to 1999, Explorer drivers are nearly twice as likely to die in rollovers as are occupants of Jeep Cherokees and Grand Cherokees, the only popular sport utilities long built like cars (Bradsher, 2000). As part of the professional responsibility of technical competence, should a senior light truck designer be aware of this rollover history?

3. The National Highway Traffic Safety Administration's mission is to "save lives, prevent injuries, and reduce vehicle-related crashes." Is NHTSA fulfilling its mission?

4. An alternative method for managing crises is described by Surowiecki (Surowiecki, 2005). Read this article, which describes the Ford-Firestone debacle as an example of poor crisis management. How involved should engineers be in their company's crisis management?

5. According to former NHTSA administrator Joan Claybrook, lawsuits are an important part of highway traffic safety. Said Claybrook,

> The lawsuits gather information. They result in sanctioning the company—sometimes with big punitive damages, which are a deterrent. And they're very important because the auto companies make a cost-benefit calculation that they're not going fix something unless it's going to cost them more not to. . . .
>
> It's a terrible thing in terms of public policy. And so these lawsuits force them—prematurely, before they want to—to redesign, fix and recall some of these vehicles or other products, and also to disclose information. (Frontline, 2002c)

The Class Action Fairness Act of 2005 limits attorney fees and moves jurisdiction from local to federal district courts (USC, 2005) in order to prevent excessive judgments. Do you believe the Ford judgments were excessive? Discuss.

Chapter 12

2003: Columbia Space Shuttle Explosion

THE REPORTED STORY

The *New York Times* Abstract:

Space shuttle Columbia breaks up on re-entry to earth's atmosphere, killing all seven astronauts aboard: Col Rick D Husband, mission commander, Capt David M Brown, Dr Kalpana Chawla, Cmdr William C McCool, Lt Col Michael P Anderson, Dr. Laurel Salton Clark and Col Ilan Ramon, an Israeli; breakup occurs 40 miles above Earth and only minutes before scheduled landing at Kennedy Space Center in Florida; shower of fiery debris falls across Texas and Louisiana; NASA will activate board of independent outside experts, led by Harold W Gehman, to oversee parts of investigation; how large a setback the loss of Columbia will pose for shuttle is difficult to assess. (Sanger, 2003)

THE BACK STORY

The Columbia was the first of the original four orbiters launched. Between its first launch in 1981 and final launch (mission STS-107) on January 16, 2003, it went through numerous upgrades, including a glass cockpit and second-generation main engines. However, more than 44% of its tiles and 41% of the 44 wing leading edge reinforced carbon-carbon panels were original equipment.

EXTERNAL TANK INSULATION

The shuttle system design, consisting of a reusable orbiter, an expendable external fuel tank carrying liquid propellants for the orbiters' engines, and two recoverable solid rocket boosters, is discussed extensively in Chapter 5. In this section, we highlight the design of the external tank insulation (Figure 12.1).

The external tank is built by Lockheed Martin. It is coated with two insulation materials: dense composite ablators for dissipating heat and low-density closed-cell foams for high insulation efficiency. Closed-cell materials consist of small pores filled with air and blowing agents that are separated by thin membranes of the foam's polymeric component. The insulation maintains an interior temperature that keeps the oxygen and hydrogen in a liquid state and an external temperature high enough to prevent ice and frost from forming on the surface.

Metallic sections that will be insulated with foam are first coated with an epoxy primer. In areas such as the bipod hand-sculpted regions, foam is directly applied over ablator materials. Where foam is applied over cured or dried foam, a bonding enhancer called Conathane is first applied to aid in adhesion. After foam is applied in the intertank region, the larger areas of foam coverage are machined down to a thickness of approximately 1 inch.

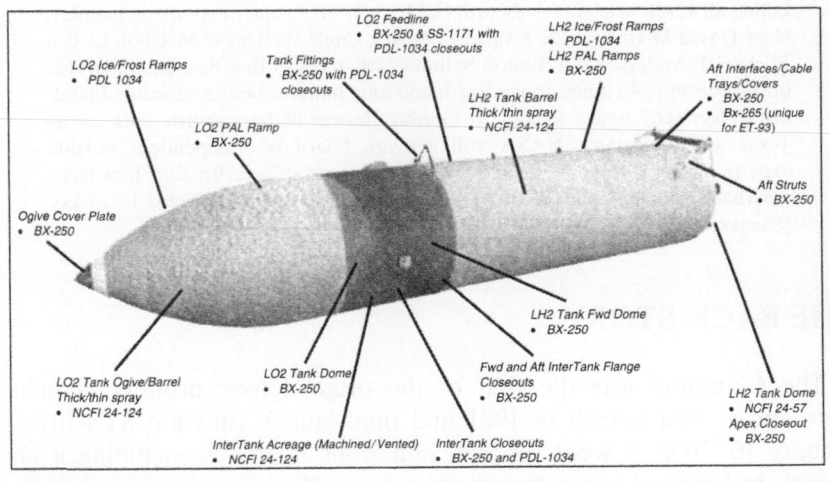

Figure 12.1 Locations of the various foam systems used on ET-93, the external tank used during Columbia's final flight.
Reprinted from NASA, 2003.

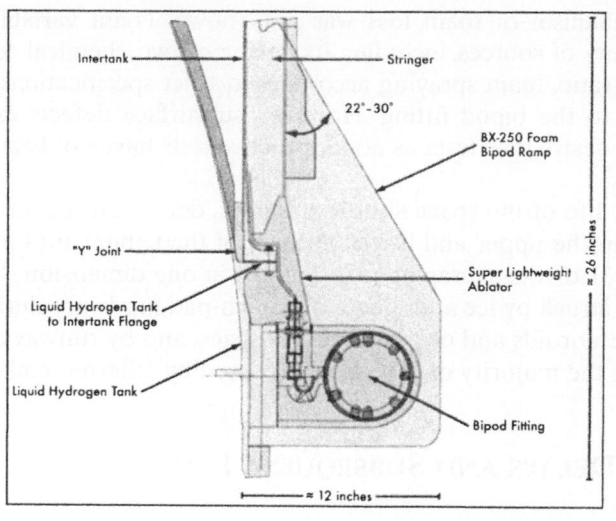

Figure 12.2 Cutaway drawing of the bipod ramp and its associated fittings. Reprinted from NASA, 2003.

Because foam applied over the bipod fittings may not provide enough protection from the high heating of exposed surfaces during ascent, the bipod fittings are coated with ablators. The bipod is one of the main connections between the external tank and orbiter. Foam is sprayed by hand over the fittings, allowed to dry, and manually shaved into a ramp shape (Figure 12.2). Only visual inspections of the bipod at the assembly facility and at the Kennedy Space Center are conducted.

EARLY PROBLEMS

Originally the bipod foam ramps on external tanks 1 through 13 possessed a 45-degree ramp angle. However, after foam was lost from the bipod ramp on mission STS-7, subsequent wind tunnel testing showed that shallower angles were aerodynamically preferable. This result caused ramp angle modification to between 22 and 30 degrees. For tanks 76 and later, a slight modification was also made to the ramp impingement profile.

Foam loss occurred on more than 80% of the 79 missions for which imagery was available. Foam was specifically lost from the left bipod ramp on nearly 10% of missions where the left bipod ramp was visible following external tank separation. For about 30% of all missions, foam loss could not be determined either because of night launch or because the external tank bipod ramp area was not in view when images were taken.

The mechanism of foam loss was not known. Foam variability comes from a variety of sources, including its mixing of two chemical components in an exact ratio, foam spraying according to strict specifications, and hand application to the bipod fitting. However, subsurface defects can only be detected by destructive tests, as nondestructive tests have not been perfected or qualified.

Over the life of the space shuttle program, debris caused an average of 143 divots in the upper and lower surfaces of the orbiter tiles during each flight, with 31 divots averaging over an inch in one dimension. Though the orbiter was struck by ice and pieces of launch-pad hardware during launch, by micrometeoroids and orbital debris in space, and by runway debris during landing, the majority of divots were caused by external tank foam loss.

LAUNCH DELAYS AND SUBSEQUENT LAUNCH

Mission STS-107 was scheduled for launch on January 11, 2001. The launch was delayed 13 times over 2 years, mainly because other missions took priority. Further, an earth-observing satellite was replaced with the FREESTAR payload, depot-level maintenance took 6 months longer than originally planned, and cracks in the main engine propellant system flow-liner were fixed. STS-107 finally launched on January 16, 2003. The assigned crew included six career astronauts and Israeli payload specialist Ilan Ramon. The launch was successful, and planned experiments took place over the next 16 days.

Columbia began its planned de-orbit and re-entry on February 1, 2003. Entry Interface (EI), arbitrarily defined as the point at which the orbiter enters the discernable atmosphere at 400,000 feet, occurred at 8:44:09 AM. At EI + 831 seconds, Columbia crossed from New Mexico into Texas and shed a wing tile. At EI + 906 seconds, the crew was informed that unusual sensor readings were being recorded and evaluated. At EI + 923 seconds, a broken response from the mission commander onboard was recorded, just as the orbiter began to disintegrate.

COLUMBIA ACCIDENT INVESTIGATION BOARD

Per changes instituted after the Challenger explosion, an external investigating board composed of seven members was automatically activated to uncover the causes of the shuttle mishap. The board named Admiral Harold Gehman, Jr. as its chair and also named five additional members. The National Aeronautics Space Administration (NASA) mobilized hundreds of personnel to directly support the board's investigation on a full-time basis. NASA officials also impounded data, software, hardware, and facilities at NASA and contractor sites, in order to ensure that all material associated with Columbia's mission was preserved as evidence.

With the assistance of the Department of Justice's Civil Division, the board established its own secure and independent database server. This server provided access to four document databases:

1. All data in NASA's process-based mission assurance system
2. All documents gathered or generated by board members, investigators, and support staff
3. All transcriptions of privileged witness interviews
4. Text of approved board meeting minutes.

Using fault tree analysis to evaluate every known factor that could have caused or contributed to the accident, more than 1000 items were considered and closed.

Based on photographic evidence, sensor data, aerodynamic/thermodynamic analyses, debris examination, and impact tests, the physical cause of the Columbia explosion was determined to be foam loss. At 81.7 seconds after launch, a piece of insulating foam separated from the left bipod ramp section of the external tank and struck the left wing in the vicinity of the lower half of reinforced carbon-carbon panel number 8. During re-entry, this breach in the thermal protection system allowed superheated air to penetrate through the leading edge insulation. This caused progressive melting of the aluminum structure of the left wing, failure of the wing, and breakup of the Columbia orbiter. The organizational cause of the explosion was a culture that relied on past success as a substitute for sound engineering practices and that prevented effective communication of critical safety information and stifled professional differences of opinion.

INVESTIGATION BOARD RECOMMENDATIONS

During the course of the board's investigation, board member Dr. Sally Ride, who also served on the Rogers commission, observed that there were "echoes" of Challenger in the Columbia explosion. (Dr. Ride was the first American female astronaut in space.) In both cases, engineers warned about an impending disaster, with NASA managers deciding to move forward. Moreover, O-ring erosion and foam loss were chronic anomalies that came to be tolerated, rather than corrected (NASA, 2003). The observation by physics Nobel prize winner Dr. Richard Feynman about shuttle safety during the Rogers commission still rang true: "When playing Russian roulette, the fact that the first shot got off safely is little comfort for the next" (Feynman, 1986). In NASA's conflicting goals of cost, schedule, and safety, safety came in third.

The board made 29 return-to-flight recommendations. The topics covered by these recommendations included the thermal protection system, imaging of the flight, orbiter sensor data, wiring, hardware bolt catchers, foam hand-spraying procedure, debris, scheduling, training, organization, recertification, and photo/drawing documentation (NASA, 2003). A recurrent theme in the recommendations was required correction of flawed practices embedded in NASA's organizational system that contributed to both explosions.

NASA's original schedule for the first space shuttle flight after the Columbia disaster was postponed from May to July 2005. Discovery flew on July 22, 2005, even though a NASA panel determined that NASA had not fully met the three most challenging of the board's recommendations. The unresolved issues include development of tile and panel repair kits and measures to eliminate all shedding of debris from the external fuel tank (Schwartz, 2005). During the July 22 launch, insulating foam again separated from the external tank.

APPLICABLE REGULATIONS

During the early days of the space shuttle program, foam loss was considered dangerous. Design engineers were extremely concerned about potential damage to the orbiter and its fragile thermal protection system, as reflected in shuttle system and external tank requirement specifications (NASA, 2003):

3.2.1.2.14 Debris Prevention: The Space Shuttle System, including the ground systems, shall be designed to preclude the shedding of ice and/or other debris from the Shuttle elements during prelaunch and flight operations that would jeopardize the flight crew, vehicle, mission success, or would adversely impact turnaround operations. (NASA, 1973)

3.2.1.1.17 External Tank Debris Limits: No debris shall emanate from the critical zone of the External Tank on the launch pad or during ascent except for such material that may result from normal thermal protection system recession due to ascent heating. (NASA, 1980)

Unfortunately, even with these specifications in place, the inaugural 1981 flight of the Columbia sustained damage from debris, causing 300 tiles to be replaced.

AN ENGINEERING PERSPECTIVE

Rodney Rocha was chief engineer in the shuttle structural engineering division. When, during a phone call on Friday, January 17 (the day after liftoff), he learned of loose foam striking the left wing of the shuttle, he gasped. All weekend, he watched a video loop showing the strike. Five days after liftoff, he was one of 30 engineers from NASA and its contractors holding the first formal meeting to assess potential damage. However, after replaying the video several times, they could not determine the strike's severity because of the camera angle. So Rocha was elected to ask shuttle mission managers for access, perhaps from American spy satellites, of images of the impacted wing area.

Two other engineers also made similar requests. All requests were denied. Manager of the shuttle engineering office Paul Shack e-mailed Rocha, "I'm not going to be Chicken Little about this." Rocha's e-mail response back to Shack included the sentence "In my humble technical opinion, this is the wrong (and bordering on irresponsible) answer."

Rocha continued to discuss his concerns with other colleagues, including Calvin Schomburg. Schomburg was an expert on heat-resisting tiles, who believed that because other foam strikes were inconsequential on previous flights, the current flight must also be safe. Schomburg's opinion had been reassuring to shuttle managers (Glanz, 2003). Beginning the weekend after the launch, a group of Boeing engineers began a mathematical analysis of the strike, working to estimate the foam debris size and its strike damage. When this inexperienced group of mathematical modelers incorrectly concluded that there no "safety of flight" risk (NASA, 2003), Rocha decided to go along with this decision because he "just wasn't being supported" and he "had faith in the abilities of our team." However, he continued to feel anxious about the strike. Rocha watched in horror from the engineering monitoring center as the shuttle landing went awry on February 1 (Glanz, 2003).

REFERENCES

Feynman, R. P., Personal observations on the reliability of the shuttle. In *Report of the Presidential Commission on the Space Shuttle Challenger Accident*, vol. 2, appendix F. Washington, D.C.: U.S. Government Printing Office, 1986. http://history.nasa.gov/rogersrep/genindex.htm.

Glanz, J., and Schwartz, J., Dogged engineer's effort to assess shuttle damage: a record of requests by experts, all rejected. *NY Times*, A1, September 26, 2003.

National Aeronautics Space Administration, *Report of the Columbia Accident Investigation Board*. Washington, D.C.: U.S. Government Printing Office, 2003. http://www.nasa.gov/columbia/home/CAIB_Vol1.html.

National Aeronautics Space Administration National Space Transportation System, *Space Shuttle Program Description and Requirements Baseline*. NSTS-0700, vol. 1, book 1. Houston, TX: NASA, 1973.

National Aeronautics Space Administration National Space Transportation System, *External Tank Component End Item Specification*. NAS8-30300. Houston, TX: NASA, 1980.

Sanger, D. E., Loss of the shuttle: the overview: shuttle breaks up, 7 Dead. *NY Times*, A1, February 2, 2003.

Schwartz, J., After 2 years on ground, shuttle is set to fly in July. *NY Times*, A10, July 1, 2005.

QUESTIONS FOR DISCUSSION

1. Of the eight ethical dilemmas presented in Chapter 1, which were present in the events leading to the Columbia explosion and its aftermath?

2. The basic shuttle system design is more than 30 years old. During its last flight, the majority of the Columbia orbiter instrumentation was 22 years old. It had been in service twice as long as its specified service life, with many sensors already failing. Of the 181 sensors in the Columbia's wings, 55 had already failed or were producing questionable readings before the final launch (NASA, 2003). Even if the flight had been successful, should the Columbia have continued to be in service?

3. Because the space shuttle program has been NASA's single most expensive activity for the past 30 years, it has been hardest hit by the NASA budget constraints of the last decade. In 1993, NASA received $4 billion; in 2002, NASA received $3.3 billion. This budget reduction had an even more severe effect on the space shuttle program because of the high priority after 1993 to complete the costly international space station. In 1993 the total space shuttle program workforce consisted of 30,391 workers; in 2002 the total space shuttle program workforce consisted of 17,462 workers (NASA, 2003). Have these budget and worker cuts affected safety? If so, how?

4. After the Challenger explosion, the space shuttles stopped launching commercial satellites and the Department of Defense began to launch all future military payloads on expendable launch vehicles (NASA, 2003). Conduct a cost-benefit analysis regarding whether the space shuttles should continue to fly.

5. What types of organizational changes are needed to change the seeming NASA overemphasis of schedule over safety?

Chapter 13

2003: Guidant Ancure Endograft System

THE REPORTED STORY

The *New York Times* Abstract:

Division of Guidant Corp, one of country's largest makers of medical devices, pleads guilty to 10 felonies, admitting it lied to government and hid thousands of serious health problems, including 12 deaths, caused by one of its products; case against division, Endovascular Technologies, results in $92.4 million in criminal and civil penalties, largest ever imposed against maker of medical devices for failing to report problems to government; company developed stent-grafts that could be used to treat abdominal aortic aneurysms without major surgery; problem was with device used to insert stent-graft; equipment could become lodged, potentially requiring emergency surgery to remove it; in some cases, it was broken into pieces before being removed; Guidant hid results that its product failed to work properly about one of every three times it was used; as part of plea, Endovascular Technologies agrees to cooperate in investigations against executives who might have been involved in wrongdoing; company also faces lawsuits from individuals. (Eichenwald, 2003)

THE BACK STORY

FDA GOOD MANUFACTURING PRACTICES

Medical devices are regulated in the United States by the Food and Drug Administration (FDA). Before a new device can be sold, it must receive FDA approval. For a Class III device, which is designed to support or sustain human life or prevent impairment of human health, its device manufacturer typically submits a premarket approval (PMA) application. A PMA application contains a device description and nonclinical laboratory studies on biocompatibility, stress, and wear. It also contains a clinical investigation section with human safety and effectiveness data, device failure and replacement data, patient complaints, and results of statistical analysis. A key component of a PMA application is a copy of all proposed labeling, including information for use and instructions for use.

A PMA application is subjected to FDA's required process of scientific review to assure device safety and effectiveness, which typically takes longer than 180 days. Upon approval, postapproval requirements such as continued safety evaluations may be imposed. For continued PMA approval, an annual postapproval report must be submitted to FDA. If a device or device malfunction may have caused or contributed to a death or serious injury, a medical device report (MDR) must be filed within 30 days of becoming aware of the event. If a device failure necessitates a labeling, manufacturing, or device modification, a PMA supplement must be submitted. These post-market surveillance procedures were implemented as a direct result of the Bjork-Shiley heart valve defect (see Chapter 8).

MEDICAL DEVICE BACKGROUND

Guidant Corporation's wholly owned subsidiary, Endovascular Technologies, Inc. (EVT), was located in Menlo Park, California. EVT manufactured the Class III Ancure Endograft System, a treatment for abdominal aortic aneurysms. An aneurysm is a diseased or weakened section of an artery wall that tends to bulge because of hardening or general deterioration of the arteries. The aorta is the largest artery in the body and transports oxygenated blood out of the heart. An abdominal aortic aneurysm (AAA) is a widening of the aorta in the area of the abdomen (Figure 13.1).

Traditionally, when an AAA diameter increases to greater than 5 centimeters and the patient is able to tolerate surgery, surgical repair is performed to

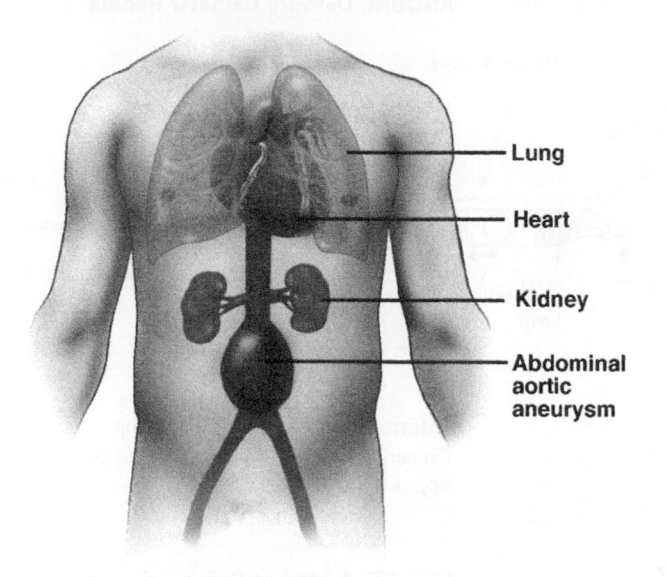

Figure 13.1   Abdominal aortic aneurysm.

prevent AAA rupture. An incision is made in the abdomen, the intestines are moved aside, the damaged portion of the aorta is removed, and a polyester patch is sewn in its place. After this surgery, the patient requires 6 months to recover.

A minimally invasive procedure to implant this patch, or endograft, in the aorta is preferable to surgical repair. The Ancure device consisted of a graft housed within a delivery catheter, or tube. The catheter was inserted in a vein in the groin and then moved to the AAA. The delivery catheter included a balloon catheter that, when inflated, floated through the blood and helped guide the device over a guidewire and secure the attachment system for the graft (Figure 13.2). Once the graft was uncompressed because its jacket was retracted, angled metal hooks and a self-expanding cylindrical metal frame attached the graft to the AAA (FDA, 1999). The delivery catheter was then removed from the body. Blood then flowed through the graft, avoiding the aneurysm, which typically shrank over time.

In the initial clinical studies submitted in the PMA application, there were 9 deaths out of 510 patients, for a 1.8% rate of device deaths, compared

ANCURE Delivery Catheter Handle

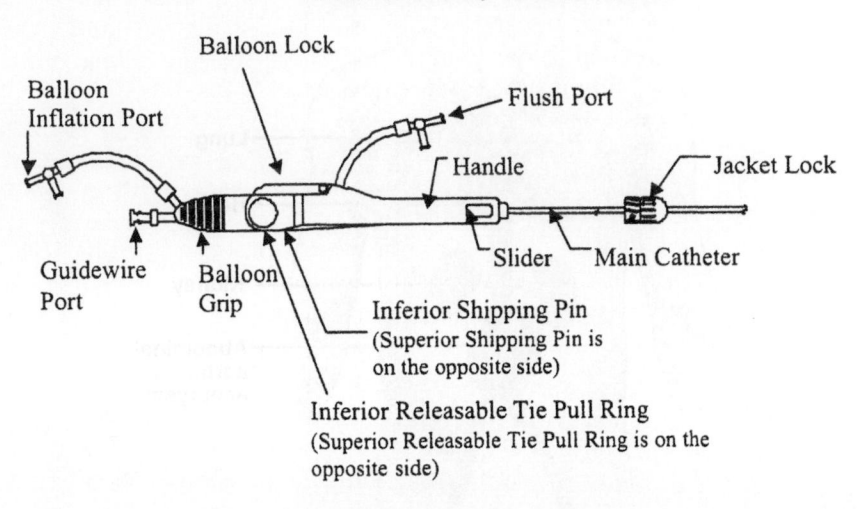

ANCURE Aortoiliac Delivery System

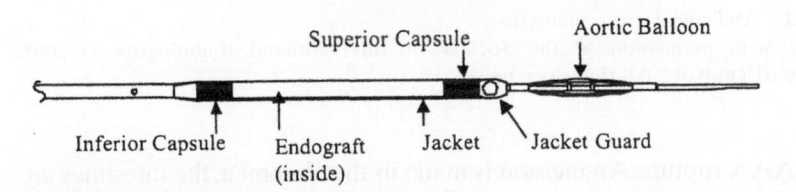

Figure 13.2   Guidant Ancure Endograft System.
Reprinted from FDA, 1999.

with the control AAA surgery, which resulted in 3 deaths out of 111 patients, or a rate of 2.7% deaths. In 4 out of 510 device cases (1%), complications such as failure to access the aneurysm or to accurately place the graft caused conversion to standard surgery (FDA, 1999).

CLINICAL PERFORMANCE AFTER FDA APPROVAL

FDA approved the Ancure device for sale in the United States on September 30, 1999. The conditions of approval included performance of a long-term follow-up study and sales case support for each device case performed (FDA, 1999). Each Ancure device cost approximately $10,000. The Ancure device received approval on the same day as its major competitor device, which was manufactured by Medtronic and considered easier to use. According to court records, some EVT officials believed that if the Ancure device could not be successfully used in a significant number of cases, it had the potential to harm marketing efforts and discourage physician customers from choosing the Ancure device.

Almost immediately after market release, the rate of Ancure device complications exceeded that of the clinical trial. In some cases, physicians were unable to implant the Ancure device because of a problem in the delivery system. In other cases, physicians were able to implant the graft, but not in a way consistent with the approved instructions for use. Some malfunctions caused the delivery system to become improperly lodged in the body, forcing some patients to undergo traditional surgical repair to remove the delivery device and correct the aneurysm (Criminal, 2003). For the total 7632 Ancure devices eventually marketed, 2800 (37%) MDRs concerning the Ancure delivery system were generated (Plea, 2003).

Some sales representatives provided information to physicians on a procedure that involved breaking or cutting the handle of the Ancure device when the delivery device became lodged and could not be removed without traditional surgery. This procedure, which became known as the "handle breaking technique," was devised in part by an EVT sales representative. It had not been pretested or presented during physician training. On January 26, 2000, the handle breaking technique was utilized unsuccessfully, and the patient undergoing that operation died (Criminal, 2003).

DEVICE INVESTIGATIONS

During a routine FDA inspection of EVT in July 2000, the inspector requested a list of all complaints regarding difficulties in catheter jacket retraction. Fifty-five complaints were provided, although more than 200 incidents existed at the time (Criminal, 2003). Later in 2000, during its own internal investigation, EVT determined that it had serious quality system requirement violations, incomplete and untimely complaint handling and documentation, incomplete MDR reporting, and inadequate correction and preventative action activities. EVT also determined that it had incomplete record keeping for process changes, poor record keeping, poor traceability practice, and significant incompliance with FDA requirements and its own internal policies (Plea, 2003). However, the results of this investigation were not reported to FDA until EVT learned it was under criminal investigation by the government (Jacobs, 2003).

EVT suspended commercial sales of the Ancure device as of March 16, 2001. On March 23, 2001, EVT disclosed to FDA the existence of 2628 additional MDRs (only 172 had been filed) concerning the Ancure device delivery system not previously reported to FDA, as required by law. Among these MDRs were 12 deaths and 57 conversions to traditional surgical repair. EVT also disclosed that it had failed to seek prior approval to amend its instructions for use to include the handle breaking technique as legally required (Plea, 2003). After revising physician instructions for troubleshooting in a PMA supplement, EVT was allowed to remarket the Ancure device in August 2001 (Jacobs, 2003).

SETTLEMENT

After 3 years of criminal investigation, the U.S. Attorney charged Guidant with 10 federal felonies. On June 12, 2003, Guidant pled guilty and agreed to a criminal fine of $32.5 million for these 10 felony violations. It also agreed to forfeit $10.9 million to compensate for profit made from illegal sales. Further, Guidant agreed to pay a civil settlement of $49 million (Plea, 2003) to settle claims that the firm's actions caused the Medicare, Medicaid, and Veteran Affairs programs to pay millions of dollars for the adulterated and misbranded devices (U.S. Attorney, 2003). As of March 2004, this total settlement of $92.4 million is the largest pay-out to date ever levied for violating FDA's medical device reporting requirements.

Guidant stopped manufacturing the Ancure device and closed its EVT facility in 2003 (Jacobs, 2003).

APPLICABLE REGULATIONS

The United States Attorney charged Guidant with 10 felony counts against the United States Code (U.S.C.):

1. Guidant was charged with one count against 18U.S.C.§1001, false statement within the jurisdiction of a federal agency, because it did not provide a full list of jacket retraction complaints to an FDA investigator when requested (Criminal, 2003):

   Title 18 Crimes and Criminal Procedure
   Sec. 1001. Statements or entries generally
   (a) Except as otherwise provided in this section, whoever, in any matter within the jurisdiction of the executive, legislative, or judicial branch of the Government of the United States, knowingly and willfully
       (1) falsifies, conceals, or covers up by any trick, scheme, or device a material fact;

    (2) makes any materially false, fictitious, or fraudulent statement or representation; or

    (3) makes or uses any false writing or document knowing the same to contain any materially false, fictitious, or fraudulent statement or entry;
    shall be fined under this title or imprisoned not more than 5 years, or both.

(b) Subsection (a) does not apply to a party to a judicial proceeding, or that party's counsel, for statements, representations, writings or documents submitted by such party or counsel to a judge or magistrate in that proceeding.

(c) With respect to any matter within the jurisdiction of the legislative branch, subsection (a) shall apply only to

    (1) administrative matters, including a claim for payment, a matter related to the procurement of property or services, personnel or employment practices, or support services, or a document required by law, rule, or regulation to be submitted to the Congress or any office or officer within the legislative branch; or

    (2) any investigation or review, conducted pursuant to the authority of any committee, subcommittee, commission or office of the Congress, consistent with applicable rules of the House or Senate. (USC, 2004)

2. Guidant was also charged with nine counts against 21U.S.C.§331(a) & §333(a)(2), interstate shipment of misbranded devices, because shipped devices were misbranded when MDRs were not filed within 30 days of malfunction and because updated, approved instructions for use were not included (Plea, 2003; U.S. Attorney, 2003):

Title 21 Food and Drugs
Sec. 331. Prohibited acts
The following acts and the causing thereof are prohibited:

(a) The introduction or delivery for introduction into interstate commerce of any food, drug, device, or cosmetic that is adulterated or misbranded.

Sec. 333. Penalties

(a) Violation of section 331 of this title; second violation; intent to defraud or mislead

(1) Any person who violates a provision of section 331 of this title shall be imprisoned for not more than one year or fined not more than $1,000, or both.

(2) Notwithstanding the provisions of paragraph (1) of this section, if any person commits such a violation after a conviction of him under this section has become final, or commits such a violation with the intent to defraud or mislead, such person shall be imprisoned for not more than three years or fined not more than $10,000, or both. (USC, 2004)

AN ENGINEERING PERSPECTIVE

After initial PMA approval in 1999, word of complications quickly spread through EVT, causing several employees to question the safety of the Ancure device. Early in 2000, after a meeting of a dozen employees, one engineer e-mailed a memo to a superior proposing that the handle breaking technique be thoroughly tested, with the problem reported to FDA for review and approval. Although some testing was conducted, the technique was not reported to FDA at that time, according to court records and accounts by ex-employees.

After only 55 complaints were given to the FDA inspector in July 2000, a small group of employees decided to take action. In October 2000, seven employees, later dubbed the "Anonymous Seven" in court documents, sent an anonymous letter to Guidant's chief compliance officer in Santa Clara, California, describing their efforts to alert the company to problems through normal channels. They charged that EVT had failed to report numerous problems to FDA, and sent a copy of this letter to FDA. This letter launched the internal EVT investigation, as well as the 3-year investigation by FDA's Office of Criminal Investigations and the Federal Bureau of Investigations (FBI) (Jacobs, 2003; U.S. Attorney, 2003).

In its plea agreement, Guidant agreed to not "initiate contacts with former employees designated by the government without prior approval of the government, and, if contacted by such designated individuals, [to] notify the government of the facts and the substance of such contacts" (Plea, 2003).

REFERENCES

Criminal Information, *U.S. v. Endovascular Technologies, Inc.*, Case No. CR-02-0179 SI, U.S. District Court (N.D.CA 0609), 2003.
Eichenwald, K., Maker admits it hid problems in artery device. *NY Times*, A1, June 13, 2003.

Food and Drug Administration, *Premarket Application for ANCURE Tube System, ANCURE Bifurcated System, ANCURE Iliac Balloon Catheter. P990017.* September 28, 1999. http://www.fda.gov/cdrh/pdf/P990017b.pdf.

Jacobs, P., Medical firm's dangerous secret device's troubles were well known at Menlo Park company. *SJ Merc News*, 1A, August 3, 2003.

Plea Agreement, *U.S. v. Endovascular Technologies, Inc.*, Case No. CR-02-0179 SI, U.S. District Court (N.D.CA 0612), 2003.

United States Attorney's Office, Northern District of California, *Endovascular Press Release*, June 12, 2003. http://www.usdoj.gov/usao/can/press/html/2003_06_12_endovascular.html.

United States Code, 2004. http://www.straylight.law.cornell.edu/uscode.

QUESTIONS FOR DISCUSSION

1. How should medical device risk be assessed?

2. *Hazard analysis* refers to the systematic use of available information to identify potential sources of harm and to estimate the accompanying risk. Hazard analysis for a medical device is described by Daniel Kamm in http://www.fmeainfocentre.com/download/risk1.pdf. Construct a hazard analysis for the Ancure Endograft System.

3. What were possible causes of the difference in complication rate during clinical trials described in the PMA application and after market release?

4. When the Anonymous Seven sent their letter to the FDA, they did not initially disclose their identities. How would identity disclosure have altered the federal investigation of EVT and the resulting settlement?

5. In early 2004, the FDA began to implement a new third-party inspection program. This program addresses the need for more FDA inspectors because most medical device facilities are not inspected biennially as required. How would you ensure that device companies are following FDA regulations?

Chapter 14

2003: Northeast Blackout

THE REPORTED STORY

The *New York Times* Abstract:

> Surge of electricity to western New York and Canada touches off series of power failures and forced blackouts that leave parts of at least eight states in Northeast and Midwest without electricity; widespread failures provoke evacuation of office buildings, strand thousands of commuters and flood some hospitals with people suffering in stifling heat; grid that distributes electricity to eastern United States becomes overloaded shortly after 4 PM, tripping circuit breakers and other protective devices at generating stations from New York to Michigan; power in New York City is shut off by officials struggling to head off wider blackout; Cleveland and Detroit go dark, as do Toronto and sections of New Jersey, Penn, Conn, and Mass; hospitals and government buildings switch on backup generators to keep essential equipment operating; airports throughout affected states suffer serious disruptions, subways in New York City go out of service, and commuter trains also come to halt; officials say cause of blackout is under investigation but that terrorism does not appear to have played role; Pres Bush says electrical grid might need to be modernized. (Barron, 2003)

THE BACK STORY

THE NORTH AMERICAN INTERCONNECTION

In 1999 the National Academy of Engineering (NAE) invited 60 engineering societies to nominate the greatest engineering achievements of the 20th century. Based on these nominations, an NAE selection committee determined the top 20 achievements. Achievement number one was wide-spread electrification, which traces its roots to the work of Thomas Edison. Edison's work led to the first commercial power plant by 1882, based on direct current (DC) generation. Later power plants utilized alternating current (AC) and were based on the work of Nikola Tesla and Charles Steinmetz. AC is preferable to DC because current losses are minimized over long distances (NAE, 2005).

The North American power grid has the capacity to generate 950,000 megawatts (MW) from nearly 3500 utility organizations to more than 100 million customers and 283 million people. Electricity is produced at lower voltages (10 to 25 kilovolts [kV]) at various generators, such as nuclear and natural gas power plants. This electricity is "stepped up" to higher voltages (230 to 765 kV) for bulk transmission across 200,000 miles of transmission lines. The higher voltages reduce electricity loss from conductor heating and allow power to be shipped economically over long distances. Transmission lines are interconnected at switching stations and substations to form a network of lines and stations called a power "grid." Electricity flows through the interconnected network of transmission lines along paths of least electrical resistance. When the power arrives near a load center, it is "stepped down" to lower voltages for distribution to customers. Some larger industrial and commercial customers use intermediate voltages levels (12 to 115 kV), but most residential customers use 120 to 240 V. Although the North American power system is commonly referred to as a grid, there are actually three distinct power grids or interconnections (Figure 14.1). The

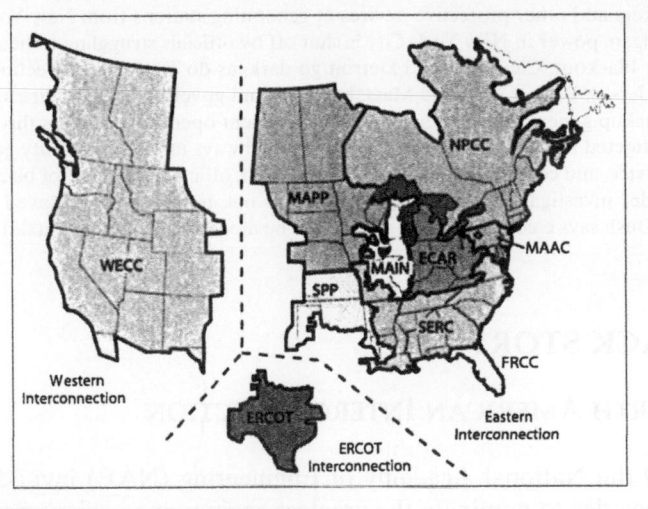

Figure 14.1 North American interconnection.
Reprinted from U.S.-Canada, 2004.

three interconnections are electrically independent from each other, except for a few small DC linkages.

GRID RELIABILITY

Grid reliability is difficult because electricity flows at close to the speed of light and is not economically storable in large quantities. Therefore electricity must be produced the instant it is used. Further, without the use of control devices prohibitively expensive for general use, AC electricity flow cannot be controlled. The North American Electric Reliability Council (NERC) and its 10 Regional Reliability Councils ensure reliability through 7 key concepts:

1. Balance power generation and demand continuously. Generator production must be scheduled or "dispatched" to meet constantly changing demands, typically on an hourly basis. Fine-tuning throughout the hour sometimes occurs through the use of automatic generation controls. Demand is somewhat predictable: highest in the summer, during the afternoon and evening, and on weekdays. Failure to match generation to demand causes the frequency of an AC power system, which is nominally 60 Hertz (Hz), to increase when generation exceeds demand and decrease when generation is less than demand.
2. Balance reactive power supply and demand to maintain scheduled voltages. Reactive power sources, such as capacitor banks and generators, must be adjusted during the day to maintain voltages within a secure range, pertaining to all system electrical equipment. Most generators have automatic voltages regulators that cause the reactive power output of generators to increase or decrease to control voltages to scheduled levels.
3. Monitor flows over transmission lines and other facilities to ensure that thermal (heating) limits are not exceeded. All lines, transformers, and other equipment carrying electricity are heated by the electricity flow through them, which must be limited to avoid overheating. Conductor heating is also affected by ambient temperature and wind. A short circuit or "flashover" can occur if an energized line gets too close to another object.
4. Keep the system in a stable condition. Voltage stability limits are set to ensure that the unplanned loss of a line or generator will not cause voltages to fall to dangerously low levels, causing point automatic relays to shed load. Power stability limits are set to

ensure that a short circuit or an unplanned loss of a line, transformer, or generator will not cause the remaining generators and loads being served to lose synchronism (60 Hz) with one another. Under extreme losses of synchronism, the grid may break apart into separate electrical islands that maintain their own frequency.

5. Operate the system so that it remains in a reliable condition even if a contingency occurs, such as the loss of a key generator or transmission facility (the "N-1 criterion"). The system must be operated at all times to ensure that it will remain in a secure condition following the loss of the most important generator or transmission facility. This loss cannot jeopardize the remaining facilities in the system to exceed emergency ratings or stability limits, which could lead to a cascading outage. If a contingency does occur, the system must be restored as soon as practical but within no more than 30 minutes to compliance with normal limits.

6. Plan, design, and maintain the system to operate reliably.

7. Prepare for emergencies. For rare events, each operating entity is required to have emergency procedures covering a credible range of emergency scenarios (U.S.-Canada, 2004).

Unfortunately, compliance with NERC policies is voluntary, rather than required by law. Grid reliability has been further hampered by the 1992 Energy Policy Act and Order 888, issued by the Federal Energy Regulatory Commission (FERC) in 1996, which deregulated the energy market. The intent of this act was to foster competition, increase efficiency, and lower energy costs. Instead, to position themselves for deregulation, utility companies tended to reorganize and sell off potentially less profitable business segments. Because mergers reduced the need for multiple engineering departments, many seasoned engineers were encouraged to retire. To increase value in units sold, some utilities reduced expenses by cutting payroll and increasing right-of-way maintenance intervals. Because it was unclear how costs incurred in the transmission network would be recovered, utility investment in the transmission networks declined (McClure, 2005).

THE GRID IN OHIO

Within the Eastern Interconnection of the North American grid, First Energy Corporation (FE) had seven subsidiary distribution utilities before the blackout: Toledo Edison, Ohio Edison, The Illuminating Company in

Ohio, and four more in Pennsylvania and New Jersey. Its Ohio control area spanned the three Ohio distribution utility footprints and that of Cleveland Public Power, a municipal utility serving the city of Cleveland. Within FE's Ohio control area was the Cleveland-Akron area, which was a transmission-constrained load pocket with relatively limited generation (U.S.-Canada, 2004).

Before the blackout, FE had 16 power plants and annual revenue of about $16 billion. The utility operated as a conglomerate, handling everything from plant operations to customer services. It benefited enormously from energy deregulation. After its Davis-Besse nuclear plant east of Toledo was shut in February 2002 for maintenance, it was discovered that boric acid ate through much of a 6-inch-thick steel cap on the plant's reactor vessel. Awaiting a Nuclear Regulatory Commission decision to reopen the plant, FE was forced to purchase power elsewhere to make up for the halt in productivity at Davis-Besse. On August 7, 2003, a federal judge ruled FE violated pollution-control laws when it rebuilt a power plant without installing state-of-the-art smog controls required under the Clean Air Act. On the day of the court's decision, Standard and Poor's held FE's credit rating one notch above junk status. In early August 2003, FE reported a second-quarter loss of $57.9 million, or 20 cents per share, as a result of special charges (Linzer, 2003).

THE INITIATION OF THE BLACKOUT

The Northeast blackout, which occurred on August 14, 2003, started with First Energy. The largest blackout in North American history, it shut down 62,000 MW of generation capacity and cost businesses an estimated $13 billion in productivity. Some 50 million users were affected over several days in eight U.S. states and Ontario, Canada (McClure, 2005).

August 14 began as a normal electric consumption day in Cleveland and Akron. During the early afternoon, moderately high loads were serving the air-conditioning demand. These loads were consuming high levels of reactive power. First Energy had previously communicated to Midwestern Independent System Operator (MISO), which balanced generation and loads in real time for that region, that two of Cleveland's active and reactive power production anchors were shut down that day (Davis-Besse and Eastlake 4). However, it did not communicate to MISO that four or five capacitor banks in the Cleveland-Akron area had been removed from service for routine inspection. These capacitor banks affected the available reactive power.

At 13:31 Eastern Daylight Time (EDT), Eastlake Unit 5, a 597-MW generating unit located west of Cleveland on Lake Erie, tripped and shut down automatically. The trip occurred when the Eastlake 5 operator sought to increase the unit's reactive power output. The attempted increase caused the unit's protection system to detect that the reactive power output exceeded its capability, causing the unit to go offline. This loss of power required FE to import additional power, making voltage management in northern Ohio more challenging. Shortly after 14:14 EDT, the alarm and logging system in FE's control room failed and was not restored until after the blackout. After 15:05 EDT, three of FE's 345-kV transmission lines began to trip because the lines were contacting overgrown trees within the lines' right-of-way areas. The first line failed because it sagged lower as it grew hotter due to heavier line loading. As each line failed, its outage increased the loading on the remaining lines.

By around 15:46 EDT, when FE, MISO, and neighboring utilities began to realize that the FE system was in jeopardy, the only way that the blackout might have been averted would have been to drop at least 1500 MW of load around Cleveland and Akron. Unfortunately, no such effort was made. By 15:46 EDT, subsequent increased loading and decreased voltage on FE's underlying 138-kV system serving Cleveland and Akron pushed those lines into overload, again because of contact of the sagging lines with other objects below the lines. Customers in Akron and areas west and south of the city lost power, causing about 600 MW of load to drop.

At 16:05:57 EDT, after 15 other lines tripped, the Sammis-Star 345-kV line tripped. This shutdown occured from protective relay action that measured low apparent impedance (depressed voltage divided by abnormally high line current). On this line, the reactive power flows were almost 10 times higher than they had been earlier in the day because of the current overload. After Sammis-Star tripped, no large-capacity transmission lines were left in southern Ohio to support the significant amount of load in northern Ohio. The Sammis-Star shutdown triggered a cascade of interruptions on the high-voltage system, causing electrical fluctuations and facility trips (Figure 14.2).

An electrical cascade is a dynamic phenomenon that cannot be stopped by human intervention once started. Within 7 minutes, the blackout rippled from the Cleveland-Akron area across much of the northeast United States and Canada. By 16:13:00 EDT, more than 508 generating units at 265 power plants were lost.

TASK FORCE INVESTIGATION AND RECOMMENDATIONS

The root causes and cascade sequence were determined through extensive investigation by the U.S.-Canada Power System Outage Task

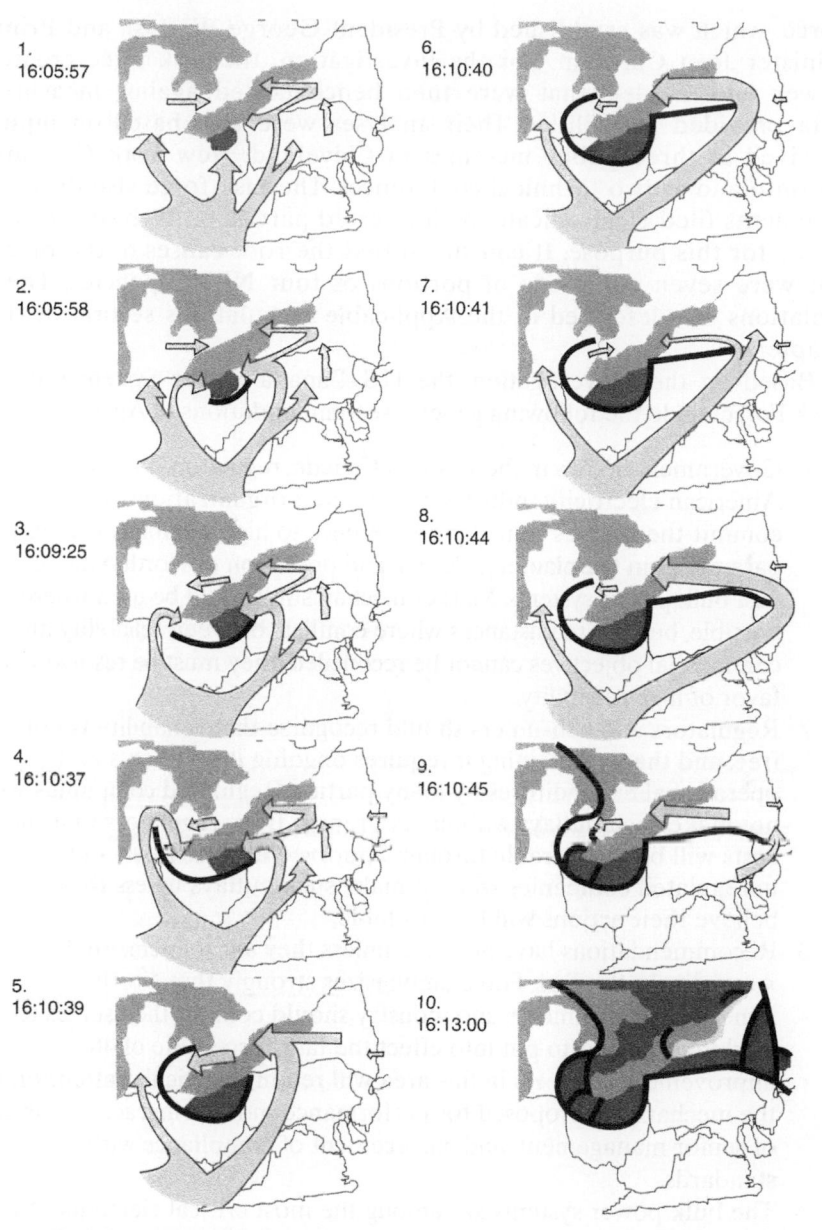

Figure 14.2 The cascade sequence. Light gray arrows represent the overall pattern of electricity flows. Black lines represent approximate points of separation between areas within the Eastern interconnect. Dark gray shading represents areas affected by the blackout. Mid-gray shading represents the Great Lakes.
Reprinted from U.S.-Canada, 2004.

Force, which was established by President George W. Bush and Prime Minister Jean Chretien. For the investigation, the task force created power flow models that were then benchmarked against measured data provided by utilities. Their analyses were also based on inputs received at three public meetings in Cleveland, New York City, and Toronto and at two technical conferences. The task force also drew on comments filed electronically by interested parties on Web sites established for this purpose. It concluded that the root causes of the blackout were seven violations of portions of four NERC policies. These violations are described in the Applicable Regulations section of this chapter.

Based on their investigation, the U.S.-Canada Power System Outage Task Force made the following general recommendations in April 2004:

1. Government bodies in the U.S. and Canada, regulators, the North American electricity industry, and related organizations should commit themselves to making adherence to high reliability standards paramount in the planning, design, and operation of North America's vast bulk power systems. Market mechanisms should be used where possible, but in circumstances where conflicts between reliability and commercial objectives cannot be reconciled, they must be resolved in favor of high reliability.

2. Regulators and consumers should recognize that reliability is not free, and that maintaining it requires ongoing investments and operational expenditures by many parties. Regulated companies will not make such outlays without assurances from regulators that the costs will be recoverable through approved electric rates, and unregulated companies will not make such outlays unless they believe their actions will be profitable.

3. Recommendations have no value unless they are implemented. Accordingly, the Task Force emphasizes strongly that North American governments and industry should commit themselves to working together to put into effect the task force suite of 46 improvements. Success in this area will require particular attention to the mechanisms proposed for performance monitoring, accountability of senior management, and enforcement of compliance with standards.

4. The bulk power systems are among the most critical elements of our economic and social infrastructure. Although the August 14 blackout was not caused by malicious acts, a number of security-related actions are needed to enhance reliability. (U.S.-Canada, 2004)

As of 2005, NERC policy compliance is still voluntary.

APPLICABLE REGULATIONS

Seven violations of portions of four NERC policies were the root causes of the blackout. These violations and policies are listed here. Note that the exact sections of policies 5 and 9 have been renumbered since the original blackout investigation because these two policies were revised in 2004.

Violation 1: Following the outage of the Chamberline-Harding 345-kV line, FE did not take the necessary actions to return the system to a safe operating state within 30 minutes.

Policy 2, Section A, Standards 1 and 2:

1. **Basic reliability requirement regarding single contingencies.** All CONTROL AREAS shall operate so that instability, uncontrolled separation, or cascading outages will not occur as a result of the most severe single contingency.
2. **Return from OPERATING SECURITY LIMIT violation.** Following a contingency or other event that results in an OPERATING SECURITY LIMIT violation, the CONTROL AREA shall return its transmission system to within OPERATING SECURITY LIMITS as soon as possible, but no longer than 30 minutes. (NERC Policy 2, 1998)

Violation 2: FE did not notify other systems of an impending system emergency.

Policy 5, Section A, Requirement 4:

1. **Information sharing.** To facilitate emergency assistance, the OPERATING AUTHORITY shall inform other potentially affected OPERATING AUTHORITIES and its RELIABILITY COORDINATOR of real time or anticipated emergency conditions, and take actions to avoid when possible, or mitigate the emergency. (NERC Policy 5, 2004)

Violation 3: FE's state estimation/contingency analysis tools were not used to assess the system conditions.

Policy 4, Section A, Requirement 5:

5. **Monitoring.** Monitoring equipment shall be used to bring to the system operator's attention important deviations in operating conditions and to indicate, if appropriate, the need for corrective action. (NERC Policy 4, 1998)

Violation 4: MISO did not notify other reliability coordinators of potential problems.

Policy 9, Section E, Requirement 1.5:

1.5. Communication with RELIABILITY COORDINATORS of potential problems. The RELIABILITY COORDINATOR who foresees a transmission problem (such as loss of reactive reserves, etc.) within its RELIABILITY COORDINATOR AREA shall issue an alert to all CONTROL AREAS and TRANSMISSION OPERATING ENTITIES in its RELIABILITY AREA, and all RELIABILITY COORDINATORS within the INTERCONNECTION via the Reliability Coordinator Information System without delay. The RELIABILITY COORDINATOR will disseminate this information to its OPERATING AUTHORITIES. (NERC Policy 9, 2004)

Violation 5: MISO was using non–real-time data to support real-time operations.

Policy 9, Section E, Requirement 1.3:

1.3. Situational awareness. The RELIABILITY COORDINATOR shall be continuously aware of conditions within its RELIABILITY COORDINATOR AREA and include this information in its reliability assessments. (NERC Policy 9, 2004)

Violation 6: PJM Interconnection and MISO as Reliability Coordinators lacked procedures or guidelines between themselves on when and how to coordinate an operating security limit violation observed by one of them in the other's area because of a contingency near their common boundary.

Violation 6 was under the jurisdiction of Policy 9, which at the time of the blackout was unspecific.

Violation 7: The monitoring equipment provided to FE operators was not sufficient to bring the operators' attention to the deviation on the system (U.S.-Canada, 2004).

Policy 4, Section A, Requirements 1 and 5:

1. **Resources.** The system operator shall be kept informed of all generation and transmission resources available for use.

5. **Monitoring.** Monitoring equipment shall be used to bring to the system operator's attention important deviations in operating conditions and to indicate, if appropriate, the need for corrective action. (NERC Policy 4, 1998)

AN ENGINEERING PERSPECTIVE

In 1996 two large-scale power outages occurred in July and August in the Pacific Northwest. Unknowingly, on August 10 the operators configured their system in a condition in which outage of the Keeler-Allston line led to cascading outages. During both the July and August failures, protection of generator field current excitation equipment failed at the McNary hydro plant. Based on a subsequent investigation by the Western System Coordinating Council, power engineers recommended substantial system changes to improve voltage support stability, to minimize power oscillation damping, and to improve simulation modeling (WSCC, 1996; Taylor, 1999).

Later, after widespread power outages in the summer of 1999, power engineers in a Department of Energy outage study team investigated root causes. Although only local causes were detailed in their report, it was generally accepted by the power industry at this time that the amount of generating capacity available during peak demand was shrinking throughout the country. Further, the transmission system was being subjected to flows in magnitudes and directions that had not been studied or for which there was minimal operating experience. Maintenance of the distribution infrastructure had suffered in recent years because utilities had been pressed for time and money. Finally, NERC at this time asked for legislative authority to make compliance with its rule-making mandatory, rather than voluntary (Sweet, 2000).

Even when the much-discounted National Energy Policy report was issued in 2001 by Vice-President Dick Cheney, it admitted the lack of capacity in the grid to satisfy demand. However, the report urged that new transmission lines be built, rather than that the existing grid become more efficient (Sweet, 2001). Based on these and other reports, when the Northeast blackout occurred on August 14, 2003, power engineers and IEEE had been sounding the alarm on potential grid failures for almost a decade (Background to the blackout, 2003).

REFERENCES

Background to the blackout: a compendium of *IEEE Spectrum* material, *IEEE Spectrum Web Special Report*, August 21, 2003. http://www.spectrum.ieee.org/WEBONLY/special/aug03/comp.html.

Barron, J., The blackout of 2003: The overview; power surge blacks out Northeast, hitting cities in 8 states and Canada; midday shutdowns disrupt millions. *NY Times*, A1, August 15, 2003.

Linzer, D., Power blackout is latest of first energy problems. *Wash Times*, B1, August 17, 2003.

McClure, G. F., Electric power transmission reliability not keeping up with conservation efforts, *IEEE Today's Engineer*, Feb, 2005.

National Academy of Engineering, *Greatest Engineering Achievements of the 20th Century*, 2005. http://www.greatachievements.org.

North America Electric Reliability Council, *Policy 2—Transmission*. Princeton, NJ: NERC, 1998. ftp://www.nerc.com/pub/sys/all_updl/oc/opman/policy2.pdf.

North America Electric Reliability Council, *Policy 4—System Coordination*. Princeton, NJ: NERC, 1998. ftp://www.nerc.com/pub/sys/all_updl/oc/opman/policy4.pdf.

North America Electric Reliability Council, *Policy 5—Emergency Operations*. Princeton, NJ: NERC, 2004. ftp://www.nerc.com/pub/sys/all_updl/oc/opman/Policy5_BOTapproved_15Jun04.pdf

North America Electric Reliability Council, *Policy 9—Reliability Coordinator Procedures*. Princeton, NJ: NERC, 2004. ftp://www.nerc.com/pub/sys/all_updl/oc/opman/Policy9_BOTapproved_15Jun04.pdf

Stier, K. J., Dirty secret: coal plants could be much cleaner. *NY Times*, B3, May 22, 2005.

Sweet, W., Restructuring the thin-stretched grid, *IEEE Spectrum*, June, 2000, 37, 43–49.

Sweet, W., and Bretz, E. A., Energy woes, *IEEE Spectrum*, July, 2001, 38, 48–53.

Taylor, C. W., Improving grid behavior, *IEEE Spectrum*, June, 1999, 36, 40–45.

U.S.-Canada Power System Outage Task Force, *Final Report on the August 14, 2003 Blackout in the United States and Canada: Causes and Recommendations*. Washington, D.C.: U.S. Government Printing Office, 2004. http://www.ieso.ca/imoweb/EmergencyPrep/blackout2003/default.asp.

Western Systems Coordinating Council, *Disturbance Report for the Power System Outage That Occurred on the Western Interconnection*. October 18, 1996. http://www.nerc.com/~filez/reports.html.

QUESTIONS FOR DISCUSSION

1. When the Exxon Valdez polluted Prince William Sound with approximately 24 million gallons of oil, Exxon was fined $150 million and settled damage claims of $900 million. Why was First Energy not fined for starting the Northeast blackout? Why did it not have to settle damage claims?

2. When the Exxon Valdez polluted Prince William Sound, Congress enacted the Oil Pollution Act of 1990 to prevent future oil spills. Why has Congress not enacted legislation to prevent another Northeast blackout?

3. View *Enron: The Smartest Guys in the Room* (Gibney, 2005), a film documentary produced by Gibney. The film was based on a book of the same name, which was written by McLean and Elkind. Energy deregulation enabled Enron to shut off power generation arbitrarily, cause rolling blackouts in California, and to manipulate electricity prices. This eventually cost California $30 billion and Governor Gray Davis his job during an election recall. As already detailed in this chapter, energy deregulation decreased reliability of the North American grid. Name some North American grid benefits of energy deregulation.

4. What ethical dilemmas were present in the First Energy system before the Northeast blackout occurred?

5. In 1996, Tampa Electric opened an innovative power plant that turned coal, the most abundant but dirtiest fossil fuel, into a relatively clean gas, which it burns to generate electricity. Besides decreasing pollution, this coal-fired power plant is also 10% more efficient than traditional coal plants. The technology used was the integrated gasification combined cycle process, which chemically strips out pollutants such as carbon dioxide from gasified coal before it is burned, rather than trying to filter it out of exhaust. However, even though this technology offers operational cost savings that offset the higher construction costs (+20%), 90% of coal plants planned in 2005 for construction are still based on old coal processing technology. According to William Fang, deputy counsel for the Edison Electric Institute, a trade association whose members account for 75% of the country's generating capacity, many of his members think that mandatory carbon controls can be kept at bay in the United States, possibly indefinitely. Mandatory carbon controls are in place in most of the world since the Kyoto Protocol came into force in February 2005 to reduce global warming (Stier, 2005). Discuss why these energy companies are resistant to change.

Chapter 15

2004: Indian Ocean Tsunami

THE REPORTED STORY

The *New York Times* Abstract:

World's most powerful earthquake in 40 years erupts underwater off Indonesian island of Sumatra, sending walls of water barreling thousands of miles and killing more than 13,000 people in half dozen countries across South and Southeast Asia; thousands more are missing or unreachable; earthquake, measuring 9.0 in magnitude, sets off tsunamis with speeds of 500 miles an hour and more, crashing into coastal areas of Sri Lanka, India, Thailand, Indonesia, Maldives and Malaysia; 40-foot-high walls of water devour everything and everyone in their paths; force is felt 3,000 miles away in Somalia, on eastern coast of Africa, where nine people are reported killed; aid agencies rush staff and equipment to region, warning that rotting bodies threaten health and water supplies; none of most affected countries have warning systems in place to detect coming onslaught and alert their citizens to move away from coastline; seismologists with United States Geological Survey say ocean west of Sumatra and island chains to its north are hot zone for earthquakes because of nonstop collision occurring there between India plate beneath Indian Ocean seabed and Burma plate under islands and that part of continent. (Waldman, 2004)

THE BACK STORY

THE SUMATRA-ANDAMAN EARTHQUAKE

When two oceanic plates collide, the younger tectonic plate rides over the edge of the older plate. This occurs because the younger plate is less dense. The older plate bends and plunges deeply into the Earth, creating a

177

trench at the plate interface. One example of this subduction zone is the plunging of the Indo-Australian plate beneath the Eurasian plate.

On December 26, 2004, this fault ruptured, allowing the edge of the Eurasian plate to spring back up. The fault slipped by as much as 50 feet in places, averaging about 33 feet of displacement along the segment off the northwestern tip of Sumatra, where the quake was centered. From the epicenter, the rupture expanded along the fault at a speed of about 1.5 miles per second toward the north-northwest, for about 720 to 780 miles. Eventually, the northern part of the fault slipped about as much as the southern part, uplifting and tilting the Andaman Islands (Figure 15.1).

Although most earthquakes last only a few seconds, this earthquake lasted about 10 minutes. The seismic magnitude of the earthquake was estimated as between 9.1 to 9.3 (NSF, 2005). The earthquake also caused sustained vibrational free oscillations at periods greater than 1000 seconds. These oscillations remained observable for weeks in broadband seismic data from global networks. The frequencies and decay rates of earth's free oscillations offered strong constraints on our planet's interior composition, mineralogy, and dynamics. Upon analysis, these data are expected to provide new perspectives on Earth's structure (Park, 2005).

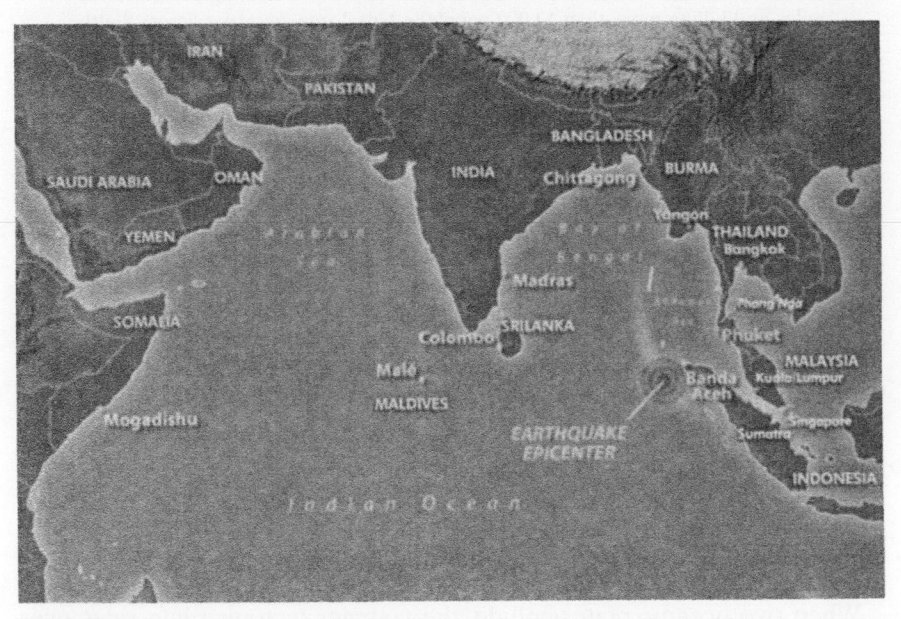

Figure 15.1 Sumatra-Andaman Earthquake Map.

THE INDIAN OCEAN TSUNAMI

The sliding of the Indo-Australian plate under its neighbor at 00:59 Greenwich Mean Time (GMT) released energy, which propagated over a wide range of frequencies and velocities. The resulting tsunami, or system of gravity waves, traveled from its seismic center near Sumatra to Africa. A tsunami is typically generated during a large earthquake of at least magnitude 7.5. In the open ocean, the waves are not visible as they travel at a velocity of 600 to 800 kilometers per hour. However, as they reach the shore, the leading edge of the wave begins to slow down while the rest of the wave grows in height. From the shore, the tsunami precursor is the recession of coastal ocean waters, with exposure of large portions of the sea floor.

At 01:02 GMT, seismic signals reached Cocos Island Station in the Indian Ocean, a facility that is part of a U.S.-led university consortium. Five minutes later, seismic signals from the quake were received at the Pacific Tsunami Warning Center (PTWC) in Hawaii from sensors in Australia, triggering an alarm. At 01:14 GMT, the PTWC issued a bulletin stating a quake of magnitude 8.0 had occurred, with no risk of tsunamis to Pacific nations. Ten minutes later, USGS revised the magnitude to 8.2.

At 01:29 GMT, Sumatra was hit by the tsunami (Figure 15.1). Thirty minutes later, PTWC issued a second bulletin upgrading the quake to magnitude 8.5 and identifying the possibility of a tsunami near the epicenter. PTWC then attempted to notify colleagues in Indonesia without success. At 2:59 GMT, eastern Sri Lanka and Thailand were hit by the tsunami. Although several satellites passing over the Bay of Bengal recorded information about tsunami wave heights, scientists did not receive the data until hours later. At 03:29 GMT, Internet newswire reports of casualties in Sri Lanka emerged as India was hit by the tsunami.

At 3:59 GMT, the U.S. ambassador in Sri Lanka set up a system to notify the prime minister in case of any large aftershocks, based on information from PTWC. The Maldives was hit by the tsunami at 4:29 GMT. At 5:24 GMT, a team at Harvard University's seismology department upgraded the status of the original quake to 8.9, using a technique that analyzes both the size and shape of seismic waves. The United States Geological Survey (USGS) was scheduled to implement this technology in 2005. At 8:14 GMT, PTWC advised the U.S. State Department about the potential threat to Madagascar and Africa. Africa was hit at approximately the same time that the warning was issued (Nature.com, 2005).

Tsunami Aftermath

The Sumatra-Andaman earthquake damaged buildings and infrastructure in the Indonesian provinces of Aceh, North Sumatra, and West Sumatra. This damage was obscured, however, by subsequent tsunami loss of life and damage. Because no tsunami alert system was in place for the Indian Ocean, devastation in coastal areas was extensive. The counts of the dead and missing exceeded 300,000 people, with 127,420 confirmed deaths and 116,368 missing in Indonesia alone. The economic damage was estimated as exceeding $13 billion (Guy Carpenter, 2005). As shown in Figure 15.2, in some coastal areas such as Khao Lak, Thailand, the tsunami razed buildings and remodeled the topography.

Unaffected countries responded to the tsunami disaster with an estimated $5 billion in aid. Physicians and nurses arrived from all over the world to assist in postdisaster relief. According to the World Health Organization, which monitors diseases such as cholera, dysentery, and malaria that are characteristic of refugee areas, these diseases were kept in check by international aid (Schiermeier, 2005).

Some people call the devastation of the livelihoods of those dependent on tourism the second tsunami. As Thai guide Jakrin Samakkee put it, "No tourists, no work, no money, big problem." Phuket, the jewel of Thai

Satellite Views of Blue Village Pakarang Resort, Khao Lak, Thailand, before (Panel A) and after (Panel B) the Tsunami. The inset map shows the affected coastline of southern Thailand.

Figure 15.2 Satellite Views of Blue Village Pakarang Resort, Kao Lak, Thailand before (Panel A) and after (Panel B) the tsunami. The inset map shows the affected coastline of southern Thailand.
Courtesy of National University of Singapore/CRISP/IKONOS.

tourism, had 111,609 international arrivals in January 2004; One year later, international arrivals fell to 13,042. Four months after the tsunami, recovery for tourist areas varied. In Thailand, 88% of hotels in Phuket were taking guests and 38% of hotels in the Phang Nga province were open as of April 1, 2005. In the Maldives, as of March 21, 2005, 70 resorts were open and 13 hotels were still under renovation. In India, which sustained minimal damage in tourist areas, all hotels are fully open. In Sri Lanka, as of March 31, 2005, 47 out of 248 hotels on the southern and eastern coasts of the island were closed for repairs. In Malaysia, which had no damage, all hotels are fully open. However, in 2005 the U.S. government was still warning against travel to the Aceh province on Sumatra because of severe damage and the ongoing threat of violence (Mydans, 2005).

TSUNAMI WARNING SYSTEMS

Tsunami deaths could have been decreased if a tsunami warning system had been in place in the Indian Ocean. An experimental warning system, Deep-ocean Assessment and Reporting of Tsunamis (DART), has been in place in the Pacific Ocean since 1999. It was deployed by the U.S. National Oceanic and Atmospheric Administration (NOAA) and links six seafloor pressure sensors to a NOAA satellite. DART is the latest component of the tsunami warning system, a cooperative venture of 26 states and countries that monitors seismic activity and tidal regimes throughout the Pacific basin (Williamson, 2005).

If a tsunami passes one of the DART sensors, the sensor registers the additional water pressure. Each $250,000 sensor can detect a rise of as little of 3 centimeters in the kilometer-high column of water above it. The sensor sends the information as acoustic chirps to a buoy on the surface, which then relays the data to NOAA's Geostationary Operational Environmental Satellite, stationed permanently above the equator. The satellite transmits the data to ground-based receiving stations. Chirp data are analyzed to determine if a tsunami's signature is present: wave speed of around 430 knots, wave front up to thousands of kilometers long, open ocean amplitude of a meter or less. Tsunami warnings themselves are issued from bases in Hawaii and Alaska.

Just 1 month after initial full operation, DART correctly predicted that a tsunami threatening Hawaii would hit its coast with just a half-meter-high wave. This prediction is credited with saving the state a coastal evacuation, estimated to cost $68 million. One month after the Indian Ocean tsunami brought this natural disaster to public awareness, the Bush administration pledged that it would add 32 more sensors in the Pacific, the Caribbean, and other waters near U.S. shores. This enhanced tsunami warning system, which will cost $37.5 million, is expected to be fully operational by mid 2007 (Williamson, 2005; Ross, 2005).

On June 30, 2005, the United Nations ocean commission agreed to work with 27 countries on a similar tsunami warning system for the Indian Ocean. The network, which is expected to operational by July 2006, is being financed by the individual countries, often with large injections of foreign aid (AP, 2005b).

APPLICABLE REGULATIONS

There are no applicable regulations.

AN ENGINEERING PERSPECTIVE

Samith Thammasaroj, a native Thai citizen, studied electrical and electronics engineering at the University of Vermont. There he changed his name to Smith. After receiving his Bachelor's degree in 1962, he returned to Thailand and joined the Thai Meteorology Service (AP, 2005a). Quickly mastering Thailand's predictable weather patterns, which revolve around wet and dry seasons, Thammasaroj decided to focus his spare energy on seismology. He began to study earthquakes and tsunamis, even though they were not considered a major problem in Thailand.

As he rose up the ranks of the meteorology department, Thammasaroj ordered staff to begin collecting earthquake data. He traveled to China to meet with seismologists. Noting that every tsunami he studied in the Pacific had been initiated by an earthquake of at least 7.4 in magnitude, Thammasaroj came to believe that a tsunami was possible locally because of pressures mounting along the region's fault lines. More disturbingly, the popular tourist area Phuket (south of Khao Lak) was in the direct path of a likely tsunami (Barta, 2005). In a letter to the Director-General of the Department of Local Administration, he presented some worst case scenario measures for official consideration. His letter also included a 10-point plan on tsunami preparation. Three of the 10 points that could have saved many lives were: (7) the need for a tsunami disaster drill, (9) public instruction on techniques that would help reduce tsunami damage, and (10) government agency creation of advance plans to deal with tsunamis. None of these recommendations was implemented.

Later, after Thammasaroj became Deputy Permanent Secretary of the now-defunct Communications Ministry, he wrote a letter of complaint to

Jadet In-Sawarng, the governor of Phuket (Rojanaphruk, 2005). That same year, he warned Thailand after an earthquake-triggered tsunami killed more than 2000 people in Papua New Guinea. This warning, made in a speech and later picked up by newspapers in the summer of 1998, quickly spread through the country, setting off panic and outrage. Villagers along the country's western coast thought the threat was imminent and ran into the hills, causing traffic accidents as they fled. Tourists checked out of their hotels (Barta, 2005). However, when no tsunami hit Thailand, furious tourism executives and government officials excoriated Thammasaroj for his judgment and forced him into retirement.

Seven years later, Thammasaroj's predictions were proven correct. Unfortunately, fearing retribution from the tourism industry and government, officials in the Thai Meteorology Department did not issue a tsunami warning. An hour before waves began hitting Thailand, they knew of the earthquake and its possible tsunami threat but remained quiet because they had no way to determine the true size of the waves. Less than 1 week after the Indian Ocean tsunami, Prime Minister Thaksin Shinawatra appointed Thammasaroj as a vice minister in charge of the newly established National Disaster Warning Office (AP, 2005a).

REFERENCES

Associated Press, Ostracized forecaster predicted tsunami. http://www.abcnews.com, January 13, 2005a.
Associated Press, UNESCO launches Indian Ocean tsunami warning system. *Hindustan Times*, A1, June 30, 2005b.
Barta, P., Thai official once reviled for tsunami prediction back in charge. *SD UT*, A4, January 11, 2005.
Guy Carpenter & Co., *Tsunami: Indian Ocean Event and Investigation into Potential Global Risks*. London: Guy Carpenter & Co., 2005.
Mydans, S., After the tsunami, rebuilding paradise. *NY Times*, 5-1, April 24, 2005.
National Science Foundation, *Analysis of the Sumatran-Andaman Earthquake Reveals the Longest Fault Rupture Ever.* June 1, 2005 Press Release. http://www.sciencedaily.com/releases/2005/05/050527104756.htm.
Nature.com, Indian Ocean Tsunami Timeline. 2005. http://www.nature.com/news/special/tsunami/index.html.
Park, J., Song, T. A., Tromp, J., Okal, E., Stein, S., Roult, G., Clevede, E., Laske, G., Kanamori, H., Davis, P., Berger, J., Braitenberg, C., Van Camp., M., Lei, X., Sun, H., Xu, H., and Rosat, S. Earth's free oscillations excited by the 26 December 2004 Sumatra-Andaman earthquake, *Science*, May 20, 2005, 308, 1139–1144.
Purdum, T. S., 'Deep Throat' unmasks himself as ex-no. 2 official at F.B.I. *NY Times*, A1, June 1, 2005.
Rojanaphruk, P., Are we doomed to report our mistakes? *The Nation (Bangkok)*, March 23, 2005.

Ross, P. E., Waiting and waiting for the next killer wave, *IEEE Spectrum*, March, 2005, 42, 17.

Schiermeier, Q., Tsunami exposes need for organized aid. *News@Nature.com*, May 6, 2005.

Waldman, W., Asia's deadly waves: disaster; thousands die as quake-spawned waves crash onto coastlines across southern Asia. *NY Times*, A1, December 27, 2004.

Williamson, M., Catch the wave, *IEE Review*, March, 2005, 51, 30–34.

QUESTIONS FOR DISCUSSION

1. What ethical dilemmas did Smith Thammasaroj experience?

2. Did the Thai Meteorology Department fulfill its professional responsibilities on December 26, 2004? Discuss.

3. Considering that only one Indian Ocean tsunami occurred in the 100 years before the 2004 tsunami, does a cost-benefit analysis justify the installation of a tsunami warning system in the Indian Ocean?

4. Do richer nations have an obligation to pay for installation of the Indian Ocean tsunami warning system when Indian Ocean nations cannot afford its cost?

5. In 2005 the identity of Deep Throat, the *Washington Post*'s secret source who helped unravel the Watergate scandal, was revealed to be W. Mark Felt. During the Watergate investigation, Felt was the number two official at the Federal Bureau of Investigation. Read about Felt in Purdum, 2005. Recall from Chapter 1 that the employee conscience is an employee working for change within his or her organization, and the observer conscience is a person working for change outside an organization. Would Felt be classified as an employee or observer conscience? Is Felt a hero or a traitor?

Part III

Individual Case Studies

In Part II we discussed national cases in which at least one engineer forewarned his or her supervisor of an impending disaster but was ignored. Using an ethical vocabulary based on previous experience and study, we discussed the decisions made by engineers and their managers that eventually made headlines.

In Part III we discuss more personal case studies. In Chapter 16, actual anonymous industrial cases of engineering ethics are presented that include an abbreviated description of each situation and how each engineer responded. Each of the 10 case studies begins with an ethics dilemma scorecard, identifying dilemmas the engineer faced in the work environment. Because the engineers in these case studies want students to be prepared for ethics situations before the students encounter them in industry, they were willing to discuss these painful events. To keep their identities secret, some small details have been changed to obscure company characteristics. Case Studies 3 and 10 are positive and are included for a balanced perspective.

Individual Case Studies

Chapter 16

Anonymous Industrial Engineering Ethics Cases

CASE 1: BIOMEDICAL ENGINEER

Ethics Dilemma Scorecard

Public Safety & Welfare	√
Data Integrity & Representation	√
Trade Secrets & Industrial Espionage	
Gift Giving & Bribery	
Principle of Informed Consent	√
Conflict of Interest	√
Accountability to Clients & Customers	√
Fair Treatment	√

TELL US YOUR STORY

After working 20 years at two other hospitals as Director of Biomedical Engineering, I moved to a similar position at a university hospital. From the outside, it looked like a great position. But during the 3 years I was there, I went through hell.

A hospital biomedical engineering department has four main areas of responsibility: preventive maintenance (PMs), equipment repair, product evaluation, and research. Preventive maintenance refers to the JCAHO-mandated testing of hospital equipment. JCAHO, or the Joint Commission on Accreditation of Healthcare Organizations, accredits health care facilities. Critical care devices, in areas such as the intensive care unit, are required to be tested for electrical safety twice per year. There were 7000 pieces of equipment in our preventive maintenance system. Each PM inspection required the technician to locate the equipment, perform the necessary inspection, and record the results. With only five technicians on my staff, they were not able to keep up with the demanding PM schedule and still perform their other responsibilities.

I tried as best as I could. The PMs of all the critical devices, like defibrillators, were always up to date, since I never wanted to put patients at risk. But PMs for less critical devices were not conducted, so about 50% of the equipment was off calibration. Shortly after being hired I realized the tremendous personnel shortage I had inherited, and began reporting this concern to upper management. Over the next 2 years I continued to report my inability to comply with JCAHO standards with my current staffing level, which included 85 pieces of filed documentation. I continued to request two additional technicians to help. These people were never hired, and it was even suggested that I falsify computer records to cover up my department's inability to complete the necessary inspections.

After 3 years, JCAHO came for their inspection. The inspector randomly pulled 30 PM numbers, and looked at their calibration records. Each piece of equipment was required to show an equipment inspection every 6 months for the past 3 years. Fifty percent of our equipment did not meet this requirement. Consequently, my department did not pass the inspection. Overall, the medical center did pass their inspection; however, the biomedical engineering department was listed as an area that needed improvement.

Immediately afterwards, I was called into the Vice President's office and asked to resign. I was accused of killing patients, which was very disturbing to me. During my tenure at this medical center, four patient-related incidents had resulted in patient death. The root causes of the incidents were user error; equipment never malfunctioned. These were a large number of incidents, compared to my experiences at other hospitals. At the time I didn't know it, but the Vice-President's goal was to outsource my department, which would result in a fixed and cheaper cost for maintenance and repairs. I refused.

The next morning, I met with my attorney, who advised me that if I stayed and was eventually fired, I could then sue because I had a strong case. But then I could be blackballed from my industry. In the end, I took a severance package and resigned. After I left, I fell into a deep depression for 2 years.

As a footnote, less than a year later, the same upper management was fired and had charges brought against them. The charges were for similar improprieties, resulting from their dealing with a couple of other departments in the medical center.

LOOKING BACK, WHAT WOULD YOU HAVE DONE DIFFERENTLY?

I performed my job well, under the circumstances I was given. There was nothing else I could do, unless I contacted the local newspaper or possibly another organization (JCAHO, Food and Drug Administration).

WHAT ARE YOU DOING NOW?

I had been teaching an undergraduate class part-time even before my resignation, and continued teaching this class. During my depression, I interviewed for hospital jobs, but never accepted the job offers I received. I was so disgusted with my field that after 2 years, I became a full-time professor in electrical engineering. I will never work in a hospital again.

CASE 2: MECHANICAL ENGINEER

Ethics Dilemma Scorecard

Public Safety & Welfare	
Data Integrity & Representation	
Trade Secrets & Industrial Espionage	√
Gift Giving & Bribery	
Principle of Informed Consent	
Conflict of Interest	
Accountability to Clients & Customers	
Fair Treatment	

TELL US YOUR STORY

It was Monday morning and many of my co-workers were busy grabbing their first cups of coffee and trading tales of their weekend exploits. The stories were never shocking or out of the ordinary because we are all engineers. But that Monday was a bit different. In the distance as I approached my desk there was a group in a manager's office having a really great time. They looked like a bunch of happy kids at Christmastime opening their "unexpected" gifts. As I joined the group my curiosity turned to bewilderment. Here were five or six engineers staring down at a manager's desk covered with what looked like wrinkled engineering notepad paper. It was the typical graph paper used by engineers everywhere, and there were notes, sketches, and calculations on the various formerly crumpled papers.

Although they looked like happy kids, what they were playing was anything but a kid's game. They were poring over someone else's work. A competitor's work at that! No permission was asked for and no permission was given to view this information. It was taken without the knowledge of our competitor. I'd heard that companies often pay someone to "dive" into a competitor's Dumpster in search of secrets, but I never ever thought I'd be witnessing it firsthand. Two thoughts immediately came to my mind. This was wrong, very wrong. What if someone such as the president of our company would walk by and catch all of us looking this stuff over? Probably grounds for dismissal, I was sure. But to my surprise, the president of the company was right in the middle of the group, congratulating the manager for his dedication.

I had difficulty hiding my distaste and disgust for the whole sorry situation and had to leave the manager's office quickly. Our company had committed an act of a common criminal in my opinion; never mind that Dumpster diving can be legal. The Dumpster-diving manager was proud of his take, and was actually being congratulated by the company president!

LOOKING BACK, WHAT WOULD YOU HAVE DONE DIFFERENTLY?

Because this behavior was accepted by the company president, there was nothing I could have done. I said nothing then, and would still do the same.

WHAT ARE YOU DOING NOW?

This happened my first or second week at the company. I worked there for 4 years. I left in search of a better work environment after many disagreements with my direct supervisor.

CASE 3: ELECTRICAL ENGINEER

Ethics Dilemma Scorecard

Public Safety & Welfare	
Data Integrity & Representation	√
Trade Secrets & Industrial Espionage	
Gift Giving & Bribery	
Principle of Informed Consent	
Conflict of Interest	
Accountability to Clients & Customers	√
Fair Treatment	

TELL US YOUR STORY

In 1974, I was working for Hewlett-Packard and wrote all the code for the HP 35 calculator. Unfortunately, my code had a round-off error, which propagated into the carry digit, which was not saved. It could be reproduced by calculating $e^{\ln 2.02}$. Instead of calculating 2.02, the result was 1.01. At the time the bug was discovered, 25,000 calculators had been shipped. I could fix the bug by changing out one of three read only memories. However, by the time the fix would be ready, 100,000 calculators would have been shipped.

A meeting was called that included engineers, salespeople, marketing, and manufacturing. When someone suggested that "Suppose we don't tell anybody," my friend heard founder David Packard's pencil break. Mr. Packard immediately stated, "Who said that? As long as my name is on the building, we are always going to be upfront with our customers." We decided to send our customers a description of the circumstances that created the bug, and let them know that when a replacement was available, they would be able to receive a replacement if requested. Only 25% of the customers ever requested the new calculator.

WHY DID YOUR COMPANY DO THE RIGHT THING?

David Packard was a strong leader who governed well and fairly. He could not have reacted in any other way.

WERE COMPANIES MORE ETHICAL IN THE 1970S THAN THEY ARE NOW?

My experience is that employees emulate the behavior of their executive management. Peer pressure caused you to do the right thing. So I believe that time is not a factor. However, I do have to admit that the times could shape employee behavior. Back then, the majority of the stock (over 90%) belonged to founders William Hewlett and David Packard. There was no pressure to increase financials every quarter as there is now.

CASE 4: GEOLOGIC ENGINEER

Ethics Dilemma Scorecard

Public Safety & Welfare	√
Data Integrity & Representation	√
Trade Secrets & Industrial Espionage	
Gift Giving & Bribery	
Principle of Informed Consent	√
Conflict of Interest	√
Accountability to Clients & Customers	√
Fair Treatment	√

TELL US YOUR STORY

During the 1980s, an elementary school was evacuated after a gasoline leak was discovered. Over 400 children were moved to alternative classrooms for several months after explosive levels of vapors were detected by the local fire department. While state officials believed that gas from a nearby storage facility tank seeped into the groundwater, the owner of the storage facility, based on tests conducted by its consultant, indicated that a gasoline station in the area could be the source of the spill. In its report, the hired consulting firm showed a map that did not indicate hydrocarbon concentrations were present in the area of the tank, even though monitor wells showed concentrations in this area. The tank was located about 600 feet north of the school, with a water-table elevation 8 feet higher than the water-table elevation of the school. The gasoline station was located about 200 feet west of the school, with a water-table elevation 2 feet higher than the water-table elevation of the school. In both consultant and state reports, groundwater was noted to flow downgradient from the tank toward the school.

I was hired by the state to simulate groundwater contaminant movement in the school area with a U.S. Geological Survey two-dimensional solute transport model widely used in groundwater contamination studies. Even before conducting this analysis, I had reviewed all the data acquired, and I agreed with my colleagues that the leaking tank was the most likely source of contamination that caused the closing of the school.

The assumptions I used as inputs to the model were based on real world conditions. Model analysis indicated that contamination leaking from the tank could have migrated to near the southern end of the school within about 16 months. In fact, about 16 months after the assumed time for the leak, contamination was first detected in monitoring wells just south of the school. My analysis implicated the tank, rather than the gasoline station, as the source of contamination, and was included in the state's report. Both reports from the state and the U.S. Environmental Protection Agency concluded that the tank was the source of contamination.

Eventually, after meeting with the governor of the state, the tank owner accepted responsibility, and cleaned up about 20,000 gallons of unleaded gasoline floating on top of shallow groundwater below the ground. Cleanup was expected to take more than 20 years. The owner also agreed to pay the costs for building a new school and for moving students to alternative classrooms until the building was completed. After a second consulting firm independently reviewed the work of the first consulting firm, the first firm lost its contract with the tank owner.

LOOKING BACK, WHAT WOULD YOU HAVE DONE DIFFERENTLY?

Nothing. The state asked me to conduct a groundwater analysis, which I conducted to the best of my ability.

WHAT ARE YOU DOING NOW?

I continue to work with the state occasionally on groundwater problems.

CASE 5: BIOMEDICAL ENGINEER

Ethics Dilemma Scorecard

Public Safety & Welfare	√
Data Integrity & Representation	√
Trade Secrets & Industrial Espionage	
Gift Giving & Bribery	
Principle of Informed Consent	√
Conflict of Interest	
Accountability to Clients & Customers	√
Fair Treatment	√

TELL US YOUR STORY

I was working at a medical device startup, reporting to the Vice President of R&D. The device we were developing attached to the wrist and was to be used in the hospital operating room. Over the course of 1 month, a mechanical engineer and I conducted many experiments on our own wrists. At the end of the month, when our wrists began to ache, I went to the biomedical library to investigate the cause of this constant pain.

Through three journal articles, I discovered that I had given myself carpal tunnel syndrome. The work the other engineer and I had been conducting involved applying pressure to our wrists that, when measured with an external pressure sensor, exceeded 200 mmHg. According to the articles, the median nerve within the carpal tunnel would be compromised if pressure within the carpal tunnel exceeded 9 mmHg below diastolic blood pressure (typically 60 mmHg). I told my boss that patients using our device might get injured. Though he thought I was exaggerating and hypothesized I might have an "unusual" wrist, I kept insisting injury was possible. Eventually, he asked me to tell our principal investigator of experiments in the operating room, an anesthesiologist, about my pain.

The anesthesiologist immediately arranged for a meeting with a vascular surgeon and hand surgeon to discuss my findings.

At this meeting, the CEO, the VP of Marketing, the VP of R&D, a mechanical engineer, and I discussed my findings with the anesthesiologist, vascular surgeon, and hand surgeon. The three physicians agreed that I had given myself carpal tunnel syndrome. Even worse, because the device was mounted on a steel wrist brace completely encircling the wrist, the two surgeons believed that during a long surgery, too little blood would circulate to the hand, causing tissue necrosis (tissue death). As we left the meeting, the VP of Marketing joked, "So you go in for hip surgery, but come out without a hand. Is this bad?!" I did not find this funny.

A few days later, we had our quarterly meeting with our technical advisor, who was also an anesthesiologist. This technical advisor was on the Board of Directors. When the technical advisor heard about carpal tunnel syndrome and tissue necrosis, he immediately mandated that the wrist brace design be changed. After the meeting, he apologized for our pain, and told the mechanical engineer and me that our company would pay for any treatment that we needed. It the first time a manager at the company had shown concern for our injuries.

LOOKING BACK, WHAT WOULD YOU HAVE DONE DIFFERENTLY?

Of course, if I had known about carpal tunnel syndrome, I would not have conducted so many experiments on myself. But I would not have changed anything else I did. I felt a duty towards not injuring patients using our device. That's why I kept complaining. I didn't know it then, but I am glad that doctors vow to "first, do no harm." Without the insistence of the anesthesiologist on the Board of Directors, the wrist brace design would not have been changed until a patient in the operating room was severely injured.

WHAT ARE YOU DOING NOW?

I left the startup less than a year later, when I got a new job at another device company. Believe it or not, this was only one of many incidents that happened to me at this startup and caused me to leave.

CASE 6: ELECTRICAL ENGINEER

Ethics Dilemma Scorecard

Public Safety & Welfare	
Data Integrity & Representation	
Trade Secrets & Industrial Espionage	
Gift Giving & Bribery	
Principle of Informed Consent	
Conflict of Interest	
Accountability to Clients & Customers	
Fair Treatment	√

TELL US YOUR STORY

In order to save money, my company decided to lay off 15% of its total work force. As a manager, I was responsible for choosing which of the four engineers in my group would be laid off. My boss advised me to choose the person least needed for specific projects slated for the next 2 years. Based on this criterion, I knew I had to fire our engineer on an H1B visa.

This was a terrible choice. The H1B was issued to my company only, and without it, the engineer would be forced to return to his unstable, impoverished country. Legally, when we terminated him, we were only required to pay for one ticket back to his homeland, but Human Resources (HR) agreed that we should issue tickets for him and the other members of his family. If the project list had been different, I could have chosen another engineer, who was younger, single, American, and could probably get a job more easily. Should I have chosen the younger engineer instead?

According to HR, because the H1B visa is issued to foreign workers whose specialized skills cannot be found among U.S. workers, it would be illegal to choose an American worker for the layoff over a foreign worker. Sympathy should not alter my choice. After an agonizing 24 hours, I chose the foreign-born engineer.

LOOKING BACK, WHAT WOULD YOU HAVE DONE DIFFERENTLY?

Nothing. I made the right choice based on company needs, not personal sympathy. Besides, as my wife pointed out, professional decisions should be made on professional needs only, not on personal needs. She pointed out that, in the past, a woman in the same job as a man would be fired first, since the man was supporting a family.

WHAT ARE YOU DOING NOW?

Well, I've never been someone who goes down with the ship. I left a few months later.

CASE 7: MECHANICAL ENGINEER

Ethics Dilemma Scorecard

Public Safety & Welfare	
Data Integrity & Representation	√
Trade Secrets & Industrial Espionage	
Gift Giving & Bribery	
Principle of Informed Consent	
Conflict of Interest	
Accountability to Clients & Customers	√
Fair Treatment	

TELL US YOUR STORY

Three months into my new job in Research and Development, I noticed that 20 to 30 large, palletized crates were sitting outside the building, next to the shipping area. My company manufactured vending machine-sized devices. Well, the devices in these crates had sat for 4 months after being sent back from Southeast Asia. The devices were sent back for upgrades, but because of other higher business priorities, the upgrades (which would have taken a few months) were not implemented. Per the terms of the contract (and my company always met the letter of the contract), the upgrades were due back in Southeast Asia in a total of 6 months. Since we didn't have enough time to complete the upgrades, and didn't want to pay penalties per the contract, we let the devices sit. At 5 months, they were shipped back to Southeast Asia, as is.

LOOKING BACK, WHAT WOULD YOU HAVE DONE DIFFERENTLY?

I would not have done anything differently. I was working on another project, and couldn't affect the situation.

WHAT ARE YOU DOING NOW?

I left the company 9 months after my start date. This situation was one of the primary reasons I left because it exemplified my company's ethics. I had no doubt that I would soon be faced with a similar situation in my project if I did not leave. All our projects were custom designs.

My company wrote long legal contracts for every project, with clauses and penalties for nonperformance. I later learned that my company had a reputation for resolving many issues through litigation. Performing to the exact wording of the contract was the first priority, with real world product performance and customer satisfaction being a secondary priority.

CASE 8: BIOMEDICAL ENGINEER

Ethics Dilemma Scorecard

Public Safety & Welfare	√
Data Integrity & Representation	√
Trade Secrets & Industrial Espionage	
Gift Giving & Bribery	
Principle of Informed Consent	
Conflict of Interest	√
Accountability to Clients & Customers	√
Fair Treatment	√

TELL US YOUR STORY

The CEO asked me to present my research data at his next Executive Committee meeting. I showed my slides to my boss, the VP of R&D, before the presentation to make sure we were in sync. What

I presented that day were the results of improvements over the course of 1 year to an old product that improved its performance statistic by 40%, a substantial improvement. After discussing how the improvements had been accomplished, I detailed how I had started a clinical study to validate this result. The clinical validation would be complete in 7 months.

The CEO and most of the VPs were impressed, and congratulated me on my work. But the VP of Marketing then pointed out that he had been responsible for publishing a paper on the old product with a performance statistic 130% over internal results, so new results would have to work out to at least a 190% improvement over internal results. Otherwise, the improvement wouldn't be enough to sell more products and raise our stock price. I replied that, based on results to date, it was impossible to reach 190%. I then said I couldn't change my clinical protocol, and my boss backed me up. To end our heated discussion, the CEO then said, "Let research do what research has to do, and then business development will take over." In the context of our discussion, that meant handing over my clinical data to Marketing once the clinical validation was complete, so Marketing could publish the data. I looked around the room, and saw no support for my position, except from my boss. Apparently, SEC fraud was not a problem for them.

Every few weeks during the clinical validation, the VP of Marketing would send me e-mail to see if the clinical validation was on schedule and if results could be available sooner. Each time, I responded that there was no way to speed up the study. Before the end of the clinical validation, I found another job.

LOOKING BACK, WHAT WOULD YOU HAVE DONE DIFFERENTLY?

Well, since this occurred in the post–Sarbannes-Oxley era, I could have filed an anonymous complaint, which would have gone directly to the Audit Committee, a subset of the Board of Directors. Even though the complaint would have been anonymous, it would have been obvious that either my boss or I complained. The Board of Directors loved my CEO. I had the feeling that they would side with the CEO, not me. And because several of the members of the Audit Committee were retired executives from the medical device industry, I would end up blackballed. It just wasn't worth it. That's why I left. And that would still be my decision today.

WHAT ARE YOU DOING NOW?

I conduct research for another device company. But I'm getting tired about always hearing about stock price. Eventually, I need to move out of public companies.

CASE 9: COMPUTER ENGINEER

Ethics Dilemma Scorecard

Public Safety & Welfare	
Data Integrity & Representation	
Trade Secrets & Industrial Espionage	
Gift Giving & Bribery	
Principle of Informed Consent	
Conflict of Interest	√
Accountability to Clients & Customers	√
Fair Treatment	√

TELL US YOUR STORY

I worked in industry for several years, and started teaching part time. I enjoyed teaching so much that I eventually became a full-time instructor. But because I have an MS, but not a PhD, I am not allowed to obtain a tenure position.

Tenure makes a lot of difference, in terms of pay. Even though I teach the same number of classes, same level of classes (juniors and seniors), and put three times as much time into my duties (besides teaching, serving on the Academic Senate and being the faculty advisor for two student organizations) as a fellow professor in the same department, he makes 175% more than I do. Admittedly, I put in all this time because I love working with my students.

Another problem with the tenure system is that tenured professors are "untouchable" and can't be fired. One tenured professor in my department receives poor evaluations from his students for his teaching and sends mass e-mails out that irritate much of the faculty, yet he is never disciplined. Another tenured professor in my department is known to be a bad lecturer, but received high student evaluations a few years ago. After an investigation was conducted, it turned out he changed the student evaluation data before turning them in! He was put on probation for a while, but is now back in department good graces.

LOOKING BACK, WHAT WOULD YOU HAVE DONE DIFFERENTLY?

If I had known that I would eventually become a full-time teacher, I would have gotten my PhD after my MS. I'm too old to do this now.

WHAT ARE YOU DOING NOW?

I continue to teach in an underpaid system and enjoy working with my students. Recently, they competed in an international competition with 300 other groups of students. Our project made it to the finalist level of 30 projects.

CASE 10: ELECTRICAL ENGINEER

Ethics Dilemma Scorecard

Public Safety & Welfare	
Data Integrity & Representation	
Trade Secrets & Industrial Espionage	√
Gift Giving & Bribery	
Principle of Informed Consent	
Conflict of Interest	
Accountability to Clients & Customers	
Fair Treatment	√

TELL US YOUR STORY

I applied for two jobs, and was offered both. On the surface, I did not think of the two companies as competitors. But during my first week at the job I accepted, I was one of several people called to a special meeting on what to do about one of our other products, which was not directly related to my new job. While this product, being a "first," had originally held the majority of market share, it had recently suffered from lack of innovation. Another company, coincidentally the other company offering me a job, had come along and improved a key feature of the product. The source of the technical improvement was unknown to everyone in the room except me. Because I had been offered the job of taking over development of this product line at the competing company, I knew the source of the improvement.

It was strange hearing people speculate as to how the other company had managed to "one-up" us, especially because we were trying to play catch up on the improvement, but our version of the improvement was not working. I did not volunteer the source of the competitive improvement, but immediately after the meeting walked into the Vice President of Engineering's office. I let him know that I had been offered a job by the competitive company, and knew their trade secret. Although I had never signed a nondisclosure agreement with the competitive company, I did not believe I could ethically divulge their trade secret. The Vice President agreed. During the 2 years I worked at my job, no one ever pressured me to divulge the trade secret. My company was never able to recapture lost market share of this product.

WHY DID YOUR COMPANY DO THE RIGHT THING?

I have no idea. Thinking back on the situation, I am surprised, first of all, that I immediately told the VP, and, second of all, that he agreed with me! However, this company was a wholly owned subsidiary of a very large public company, so perhaps it was the company culture to do the right thing when confronted with an ethical dilemma.

WHAT ARE YOU DOING NOW?

I've moved on to other work, but this was probably the most ethical company for which I ever worked. But I didn't realize this until years later.

APPENDIX

National Society of Professional Engineers (NSPE) Code of Ethics for Engineers

PREAMBLE

Engineering is an important and learned profession. As members of this profession, engineers are expected to exhibit the highest standards of honesty and integrity. Engineering has a direct and vital impact on the quality of life for all people. Accordingly, the services provided by engineers require honesty, impartiality, fairness, and equity, and must be dedicated to the protection of the public health, safety, and welfare. Engineers must perform under a standard of professional behavior that requires adherence to the highest principles of ethical conduct.

I. FUNDAMENTAL CANONS

Engineers, in the fulfillment of their professional duties, shall:

1. Hold paramount the safety, health and welfare of the public.
2. Perform services only in areas of their competence.
3. Issue public statements only in an objective and truthful manner.
4. Act for each employer or client as faithful agents or trustees.
5. Avoid deceptive acts.
6. Conduct themselves honorably, responsibly, ethically, and lawfully so as to enhance the honor, reputation, and usefulness of the profession.

II. RULES OF PRACTICE

1. Engineers shall hold paramount the safety, health, and welfare of the public.
   a. If engineers' judgment is overruled under circumstances that endanger life or property, they shall notify their employer or client and such other authority as may be appropriate.
   b. Engineers shall approve only those engineering documents that are in conformity with applicable standards.
   c. Engineers shall not reveal facts, data, or information without the prior consent of the client or employer except as authorized or required by law or this Code.
   d. Engineers shall not permit the use of their name or associate in business ventures with any person or firm that they believe are engaged in fraudulent or dishonest enterprise.
   e. Engineers shall not aid or abet the unlawful practice of engineering by a person or firm.
   f. Engineers having knowledge of any alleged violation of this Code shall report thereon to appropriate professional bodies and, when relevant, also to public authorities, and cooperate with the proper authorities in furnishing such information or assistance as may be required.
2. Engineers shall perform services only in the areas of their competence.
   a. Engineers shall undertake assignments only when qualified by education or experience in the specific technical fields involved.
   b. Engineers shall not affix their signatures to any plans or documents dealing with subject matter in which they lack competence, nor to any plan or document not prepared under their direction and control.
   c. Engineers may accept assignments and assume responsibility for coordination of an entire project and sign and seal the engineering documents for the entire project, provided that each technical segment is signed and sealed only by the qualified engineers who prepared the segment.
3. Engineers shall issue public statements only in an objective and truthful manner.
   a. Engineers shall be objective and truthful in professional reports, statements, or testimony. They shall include all relevant and pertinent information in such reports, statements, or testimony, which should bear the date indicating when it was current.

     b. Engineers may express publicly technical opinions that are founded upon knowledge of the facts and competence in the subject matter.

     c. Engineers shall issue no statements, criticisms, or arguments on technical matters that are inspired or paid for by interested parties, unless they have prefaced their comments by explicitly identifying the interested parties on whose behalf they are speaking, and by revealing the existence of any interest the engineers may have in the matters.

4. Engineers shall act for each employer or client as faithful agents or trustees.

     a. Engineers shall disclose all known or potential conflicts of interest that could influence or appear to influence their judgment or the quality of their services.

     b. Engineers shall not accept compensation, financial or otherwise, from more than one party for services on the same project, or for services pertaining to the same project, unless the circumstances are fully disclosed and agreed to by all interested parties.

     c. Engineers shall not solicit or accept financial or other valuable consideration, directly or indirectly, from outside agents in connection with the work for which they are responsible.

     d. Engineers in public service as members, advisors, or employees of a governmental or quasi-governmental body or department shall not participate in decisions with respect to services solicited or provided by them or their organizations in private or public engineering practice.

     e. Engineers shall not solicit or accept a contract from a governmental body on which a principal or officer of their organization serves as a member.

5. Engineers shall avoid deceptive acts.

     a. Engineers shall not falsify their qualifications or permit misrepresentation of their or their associates' qualifications. They shall not misrepresent or exaggerate their responsibility in or for the subject matter of prior assignments. Brochures or other presentations incident to the solicitation of employment shall not misrepresent pertinent facts concerning employers, employees, associates, joint venturers, or past accomplishments.

     b. Engineers shall not offer, give, solicit or receive, either directly or indirectly, any contribution to influence the award of a contract by public authority, or which may be reasonably construed by the public as having the effect of intent to influencing the awarding of a contract. They shall not offer any gift or other valuable consideration in order to secure work. They shall not pay a commission, percentage, or brokerage fee in order to secure work, except to a bona fide employee or bona fide established commercial or marketing agencies retained by them.

III. PROFESSIONAL OBLIGATIONS

1. Engineers shall be guided in all their relations by the highest standards of honesty and integrity.
   a. Engineers shall acknowledge their errors and shall not distort or alter the facts.
   b. Engineers shall advise their clients or employers when they believe a project will not be successful.
   c. Engineers shall not accept outside employment to the detriment of their regular work or interest. Before accepting any outside engineering employment they will notify their employers.
   d. Engineers shall not attempt to attract an engineer from another employer by false or misleading pretenses.
   e. Engineers shall not promote their own interest at the expense of the dignity and integrity of the profession.
2. Engineers shall at all times strive to serve the public interest.
   a. Engineers shall seek opportunities to participate in civic affairs; career guidance for youths; and work for the advancement of the safety, health, and well-being of their community.
   b. Engineers shall not complete, sign, or seal plans and/or specifications that are not in conformity with applicable engineering standards. If the client or employer insists on such unprofessional conduct, they shall notify the proper authorities and withdraw from further service on the project.
   c. Engineers shall endeavor to extend public knowledge and appreciation of engineering and its achievements.
3. Engineers shall avoid all conduct or practice that deceives the public.
   a. Engineers shall avoid the use of statements containing a material misrepresentation of fact or omitting a material fact.
   b. Consistent with the foregoing, engineers may advertise for recruitment of personnel.
   c. Consistent with the foregoing, engineers may prepare articles for the lay or technical press, but such articles shall not imply credit to the author for work performed by others.
4. Engineers shall not disclose, without consent, confidential information concerning the business affairs or technical processes of

any present or former client or employer, or public body on which they serve.

   a. Engineers shall not, without the consent of all interested parties, promote or arrange for new employment or practice in connection with a specific project for which the engineer has gained particular and specialized knowledge.

   b. Engineers shall not, without the consent of all interested parties, participate in or represent an adversary interest in connection with a specific project or proceeding in which the engineer has gained particular specialized knowledge on behalf of a former client or employer.

5. Engineers shall not be influenced in their professional duties by conflicting interests.

   a. Engineers shall not accept financial or other considerations, including free engineering designs, from material or equipment suppliers for specifying their product.

   b. Engineers shall not accept commissions or allowances, directly or indirectly, from contractors or other parties dealing with clients or employers of the engineer in connection with work for which the engineer is responsible.

6. Engineers shall not attempt to obtain employment or advancement or professional engagements by untruthfully criticizing other engineers, or by other improper or questionable methods.

   a. Engineers shall not request, propose, or accept a commission on a contingent basis under circumstances in which their judgment may be compromised.

   b. Engineers in salaried positions shall accept part-time engineering work only to the extent consistent with policies of the employer and in accordance with ethical considerations.

   c. Engineers shall not, without consent, use equipment, supplies, laboratory, or office facilities of an employer to carry on outside private practice.

7. Engineers shall not attempt to injure, maliciously or falsely, directly or indirectly, the professional reputation, prospects, practice, or employment of other engineers. Engineers who believe others are guilty of unethical or illegal practice shall present such information to the proper authority for action.

   a. Engineers in private practice shall not review the work of another engineer for the same client, except with the knowledge of such engineer, or unless the connection of such engineer with the work has been terminated.

b. Engineers in governmental, industrial, or educational employ are entitled to review and evaluate the work of other engineers when so required by their employment duties.

c. Engineers in sales or industrial employ are entitled to make engineering comparisons of represented products with products of other suppliers.

8. Engineers shall accept personal responsibility for their professional activities, provided, however, that engineers may seek indemnification for services arising out of their practice for other than gross negligence, where the engineer's interests cannot otherwise be protected.

a. Engineers shall conform with state registration laws in the practice of engineering.

b. Engineers shall not use association with a nonengineer, a corporation, or partnership as a "cloak" for unethical acts.

9. Engineers shall give credit for engineering work to those to whom credit is due, and will recognize the proprietary interests of others.

a. Engineers shall, whenever possible, name the person or persons who may be individually responsible for designs, inventions, writings, or other accomplishments.

b. Engineers using designs supplied by a client recognize that the designs remain the property of the client and may not be duplicated by the engineer for others without express permission.

c. Engineers, before undertaking work for others in connection with which the engineer may make improvements, plans, designs, inventions, or other records that may justify copyrights or patents, should enter into a positive agreement regarding ownership.

d. Engineers' designs, data, records, and notes referring exclusively to an employer's work are the employer's property. The employer should indemnify the engineer for use of the information for any purpose other than the original purpose.

e. Engineers shall continue their professional development throughout their careers and should keep current in their specialty fields by engaging in professional practice, participating in continuing education courses, reading in the technical literature, and attending professional meetings and seminars.

—As Revised January 2003

"By order of the United States District Court for the District of Columbia, former Section 11(c) of the NSPE Code of Ethics prohibiting competitive bidding, and all policy statements, opinions, rulings or other guidelines interpreting its scope, have been rescinded as unlawfully interfering with the legal right of engineers, protected under the antitrust laws, to provide price information to prospective clients; accordingly, nothing contained in the NSPE Code of Ethics, policy statements, opinions, rulings or other guidelines prohibits the submission of price quotations or competitive bids for engineering services at any time or in any amount."

STATEMENT BY NSPE EXECUTIVE COMMITTEE

In order to correct misunderstandings which have been indicated in some instances since the issuance of the Supreme Court decision and the entry of the Final Judgment, it is noted that in its decision of April 25, 1978, the Supreme Court of the United States declared: "The Sherman Act does not require competitive bidding."

It is further noted that as made clear in the Supreme Court decision:

1. Engineers and firms may individually refuse to bid for engineering services.
2. Clients are not required to seek bids for engineering services.
3. Federal, state, and local laws governing procedures to procure engineering services are not affected, and remain in full force and effect.
4. State societies and local chapters are free to actively and aggressively seek legislation for professional selection and negotiation procedures by public agencies.
5. State registration board rules of professional conduct, including rules prohibiting competitive bidding for engineering services, are not affected and remain in full force and effect. State registration boards with authority to adopt rules of professional conduct may adopt rules governing procedures to obtain engineering services.
6. As noted by the Supreme Court, "nothing in the judgment prevents NSPE and its members from attempting to influence governmental action. . ."

NOTE: In regard to the question of application of the Code to corporations vis-à-vis real persons, business form or type should not negate nor influence conformance of individuals to the Code. The Code deals with professional services, which services must be performed by real persons. Real persons in turn establish and implement policies within business structures. The Code is clearly written to apply to the Engineer and items incumbent on members of NSPE to endeavor to live up to its provisions. This applies to all pertinent sections of the Code.

Revised January 2003. Reprinted with permission of NSPE.

Institute of Electrical and Electronic Engineers (IEEE) Code of Ethics

We, the members of IEEE, in recognition of the importance of our technologies in affecting the quality of life throughout the world, and in accepting a personal obligation to our profession, its members and the communities we serve, do hereby commit ourselves to the highest ethical and professional conduct and agree:

1. to accept responsibility in making engineering decisions consistent with the safety, health and welfare of the public, and to disclose promptly factors that might endanger the public or the environment;
2. to avoid real or perceived conflicts of interest whenever possible, and to disclose them to affected parties when they do exist;
3. to be honest and realistic in stating claims or estimates based on available data;
4. to reject bribery in all its forms;
5. to improve the understanding of technology, its appropriate application, and potential consequences;
6. to maintain and improve our technical competence and to undertake technological tasks for others only if qualified by training or experience, or after full disclosure of pertinent limitations;
7. to seek, accept, and offer honest criticism of technical work, to acknowledge and correct errors, and to credit properly the contributions of others;
8. to treat fairly all persons regardless of such factors as race, religion, gender, disability, age, or national origin;
9. to avoid injuring others, their property, reputation, or employment by false or malicious action;
10. to assist colleagues and co-workers in their professional development and to support them in following this code of ethics.

Approved by the IEEE Board of Directors August 1990
© 2005 IEEE. Reprinted with permission of the IEEE.

SARBANES-OXLEY ACT OF 2002

SEC. 806. PROTECTION FOR EMPLOYEES OF PUBLICLY TRADED COMPANIES WHO PROVIDE EVIDENCE OF FRAUD

(a) IN GENERAL Chapter 73 of title 18, United States Code, is amended by inserting after section 1514 the following:

Sec. 1514A. Civil action to protect against retaliation in fraud cases

(a) WHISTLEBLOWER PROTECTION FOR EMPLOYEES OF PUBLICLY TRADED COMPANIES No company with a class of securities registered under section 12 of the Securities Exchange Act of 1934 (15 U.S.C. 78l), or that is required to file reports under section 15(d) of the Securities Exchange Act of 1934 (15 U.S.C. 780(d)), or any officer, employee, contractor, subcontractor, or agent of such company, may discharge, demote, suspend, threaten, harass, or in any other manner discriminate against an employee in the terms and conditions of employment because of any lawful act done by the employee—

   (1) to provide information, cause information to be provided, or otherwise assist in an investigation regarding any conduct which the employee reasonably believes constitutes a violation of section 1341, 1343, 1344, or 1348, any rule or regulation of the Securities and Exchange Commission, or any provision of Federal law relating to fraud against shareholders, when the information or assistance is provided to or the investigation is conducted by—

      (A) a Federal regulatory or law enforcement agency;

      (B) any Member of Congress or any committee of Congress; or

      (C) a person with supervisory authority over the employee (or such other person working for the employer who has the authority to investigate, discover, or terminate misconduct); or

   (2) to file, cause to be filed, testify, participate in, or otherwise assist in a proceeding filed or about to be filed (with any knowledge of the employer) relating to an alleged violation of section 1341, 1343, 1344, or 1348, any rule or regulation of the Securities and Exchange Commission, or any provision of Federal law relating to fraud against shareholders.

(b) ENFORCEMENT ACTION

   (1) IN GENERAL A person who alleges discharge or other discrimination by any person in violation of subsection (a) may seek relief under subsection (c), by—

      (A) filing a complaint with the Secretary of Labor; or

      (B) if the Secretary has not issued a final decision within 180 days of the filing of the complaint and there is no showing that such delay is due to the bad faith of the claimant, bringing an action at law or equity for de novo review in the appropriate district court of the United States, which shall have jurisdiction over such an action without regard to the amount in controversy.

   (2) PROCEDURE

      (A) IN GENERAL An action under paragraph (1)(A) shall be governed under the rules and procedures set forth in section 42121(b) of title 49, United States Code

      (B) EXCEPTION Notification made under section 42121(b)(1) of title 49, United States Code, shall be made to the person named in the complaint and to the employer.

      (C) BURDENS OF PROOF An action brought under paragraph (1)(B) shall be governed by the legal burdens of proof set forth in section 42121(b) of title 49, United States Code.

      (D) STATUTE OF LIMITATIONS An action under paragraph (1) shall be commenced not later than 90 days after the date on which the violation occurs.

(c) REMEDIES

   (1) IN GENERAL An employee prevailing in any action under subsection (b)(1) shall be entitled to all relief necessary to make the employee whole.

(2) COMPENSATORY DAMAGES Relief for any action under paragraph (1) shall include—
   (A) reinstatement with the same seniority status that the employee would have had, but for the discrimination;
   (B) the amount of back pay, with interest; and
   (C) compensation for any special damages sustained as a result of the discrimination, including litigation costs, expert witness fees, and reasonable attorney fees.
(d) RIGHTS RETAINED BY EMPLOYEE Nothing in this section shall be deemed to diminish the rights, privileges, or remedies of any employee under any Federal or State law, or under any collective bargaining agreement.

Sarbanes-Oxley Act of 2002, Sec. 806.

Index

American Institute for Steel Construction
(AISC)
  *Code of standard practice for steel
    buildings & bridges*, 60
  *Designing with Structural Steel: a Guide
    for Architects*, 67
  Manual of Steel Construction, 55
Anonymous Seven, 29, 160
Arctic National Wildlife Refuge, 87
At-will employment, 37
Automobile safety, 39–41
  Ford lifeguard design, 40
  Highway Safety Act, 40
  Motor Vehicle Safety Act, 40
  Nader, Ralph. *See* Nader, Ralph
  National Highway Traffic Safety
    Administration (NHTSA)
    History, 40, 137
    Investigation, Ford Pinto, 45–46
    Investigations, stability, 137–138
  Safety door locks, 39
  Standard 301, 41
    Passage delays, 44–45
    Text, 47–49

Bell Laboratories
  Fraud, 119
    Aftermath, 128
    Investigation, 126–128
    Warning, early, 130
  History, 120–123
Bemer, Bob, 116
Bjork, Viking, 100–101, 107
Bjork-Shiley heart valve
  Complications, disc, 102
    Movement obstruction, 102
    Rupture, 102
    Warning, early, 107
  Designs, 100–101
  Investigations, 102–104

Lawsuits, 104–105
  U.S. civil suit, 104
  Bowling v. Pfizer settlement, 104–105
Recall, 99, 104
Warning, early, 107
Boisjoly, Roger, 71–73
  O-ring investigations, 71–72
  Pre-launch meeting, 71

California statutes
  Environmental Quality Act, 95–96
    Earthquake safety, 96
  False advertising statute, 140
Case studies
  Individual, anonymous, 185–202
    Biomedical engineer, 187–189,
      194–195, 198–200
    Computer engineer, 200–201
    Electrical engineer, 191–192, 196–197,
      201–202
    Geologic engineer, 192–194
    Mechanical engineer, 189–191, 197–198
  National, 37–184
    Bell Laboratories scientific fraud,
      119–132
    Bjork-Shiley heart valve defect,
      99–108
    Challenger space shuttle explosion,
      63–74
    Columbia space shuttle explosion,
      145–152
    Exxon Valdez oil spill, 75–88
    Ford Explorer rollover, 133–144
    Ford Pinto recall, 39–52
    Guidant Ancure endograft system,
      153–162
    Indian Ocean tsunami, 177–184
    Kansas City Hyatt Regency skywalk
      collapse, 53–62
    Northeast blackout, 163–176

Case studies (*continued*)
   San Francisco-Oakland bay
      bridge earthquake collapse,
      89–98
   Y2K software conversion, 109–118
Challenger space shuttle
   Design, 63–66
   Explosion, 63, 69
   Warning, early, 71–73
   Investigations
      Boisjoly, Roger. *See* Boisjoly, Roger
      Presidential commission, 69–71
   Launch, 51-L
      Delays, 68
      Timeline, 68–69
   O-ring joint sealing
      Design
         Space shuttle, 67–68
         Titan III, 67
   Safety reporting structure, 71, 74
Coal slurry, 3
   Definition, 4
   Spill, Massey Energy, 4–6
Collapse, 87
Columbia space shuttle, 145, 152
   Design
      External tank insulation, 146–147
   Explosion, 145
      Warning, early, 151
   Foam loss, 147–148
      Specifications, 150
   Investigation, 148–149
      Recommendations, 149–150
   Launch, STS-107
      Delays, 148
      Re-entry, 148
Corapi, John, 32–33
Corporate average fuel economy (CAFE)
   standards, 134
   Sport utility vehicle (SUV) loophole, 134
Corporation, The, 20–21
Cost-benefit analysis, 40–41
   Auto safety, 40–41, 44–45
   Human life, 44, 51–52

Deep Throat, 184

Engineering ethics
   Codes, 13–14
      Effectiveness of, 14
      IEEE code of ethics, 13, 210

   NSPE code of ethics for engineers,
      13, 203–209
   Definition, 6
*Enron: The Smartest Guys in the
   Room*, 174
Ethical dilemmas
   Accountability to clients & customers, 18
   Conflict of interest, 18
   Data integrity & representation, 17
   Fair treatment, 18
   Gift giving & bribery, 17
   Principle of informed consent, 17
   Public safety & welfare, 16
   Trade secrets & industrial espionage, 17
Ethical theories, 7–13
   Duty ethics, 10–11
   Rights ethics, 11–12
   Utilitarianism, 9–10
   Virtue ethics, 12–13
Exxon Valdez
   Hazelwood, Joseph, 79–80
   Last voyage, 79–80
   Oil spill, 75, 79–80, *See also* Oil spill,
      Alaskan

False Claims Act, 30–31, 35
Fog of War, The, 51
Food and Drug Administration (FDA)
   Good manufacturing practices, 154
   Hazard analysis, 161
   Premarket approval, 154
      Medical device report, 154
Ford Explorer
   Bickerstaff, David, 141–142
   Design
      Stability defects, 136–137
      Stability early warning, 140–142
      2002 Model, 139
   Firestone tires, 133–134
   Ford Bronco II, 135
   Investigation, NHTSA, 138
   Lawsuits, 138–139
      Advertising settlement, 133
   Marketing, 135
Ford Pinto
   Design defect, 42–43
      Warning, early, 49–50
   Investigations, safety
      Dowie, Mark, 45
      Ford engineers, 49–50
      NHTSA, 45–46

Lawsuits, 46–47
Project goals & schedule, 41–42
Recall, 39, 46

Gioia, Dennis, xvii–xviii, 7
Graham, David, 34–35
Greenhouse, Bunnatine, 30
Guidant Ancure endograft system
  Complications
    Clinical study, 155–156
    Post-approval, 156–157
    Warning, early, 160
  Design, 155–156
  Graft surgery, 155
  Investigations
    FDA onsite, 157
    Internal, 157–158
    Settlement, 153, 158
    U.S. attorney, 158–160
  Recall, 157

Indian Ocean tsunami, 177
  Aftermath, 180–181
  Earthquake
    Mechanism, tsunami, 177–178
    Sumatra-Andaman, 178
  Timeline, 179
  Warning
    Systems, 181–182
    Early, 182–183
Insider, The, 34

Kansas City Hyatt Regency, 53
  Atrium roof collapse, 57
  Development process, 54–55
  Walkway
    Collapse, 53, 57
    Collapse administrative hearing, 58–59
    Collapse investigation, 57–58
    Design, general, 53–54
    Design, hanger rod, 56–57
    Design warning, 59

Methyl tertiary butyl ether (MTBE), 87–88
Mine Safety & Health Administration
  (MSHA), 4–6

Nader, Ralph, 40
  Unsafe At Any Speed, 40
Nanotechnology, 123
  Schon, J. Hendrik. *See* Schon, J. Hendrik

North American Electric Reliability Council
  (NERC) policies, 170–172
North American Interconnection, 163–165
  Design, 164–165
  National Academy of Engineering
    achievement, greatest, 163–164
  Ohio grid, 166–167
  Reliability, 165–166
    Deregulation and, 166
    Voluntary compliance, 166
Northeast blackout, 163
  Investigation, 168, 170
    Recommendations, 170
  Timeline, 167–169
  Warnings, early, 172–173

Oil Pollution Act of 1990, 82–85
Oil spill, Alaskan
  Animals affected, 81
  Area covered, 80–81
  Cleanup, 80–81
  Gallons spilled, 80
  Investigation, 81–82
    Recommendations, 81–82
  Lawsuits, 82
  Preparedness plans, 78–79
    Alyeska, 78
    Exxon, 78
    Implementation, 78–79
    National Response Team, 78
    Port Prince William Sound pollution
      action, 78
    Regional Response Teams, 78
    State of Alaska, 78
  Warning, early, 85–86
Options for action
  Departure, 23–24
  Employee conscience, 25–30
    Employee protection legislation,
      26–27
    Employee protection procedures, 28–29
  Observer conscience, 30–33
    Observer protection legislation, 30–31
    Observer protection procedures, 31–32
  Whistleblowing, 24–25

Paris airport roof collapse, 60–61
Personal engineering ethics threshold
  Determination of, 18–19
  Options for action when reached.
    See also Options for action

Physiology
  Abdominal aortic aneurysm, 154–155
  Heart valve, 99–100
Problem solving, 6
Professional responsibilities, 14–16
  Protection of public safety, 14–15
  Technical competence, 15
  Timely communication of
      negative & positive results to
      management, 15

Rocha, Rodney, 151

Safe Medical Devices Act of 1990
  Medical device report, 108
  Text, 105–107
San Francisco-Oakland Bay Bridge,
  New
      Construction, 95
      Design, 94–95, 97–98
      Engineering and design advisory
          panel, 98
  Original
      Collapse, 89, 93–94
      Construction, 91–93
      Design, 90–93
  Warning, early collapse, 96–97
Sarbanes-Oxley corporate reform act of
  2002, 3
  Anonymous reporting system, 28, 35
  Whistleblower protection clause, 27,
      211–213
Schon, J. Hendrik, 124–125
Spadaro, Jack, 4
Spiro, Larry, vi, xi

Thammasaroj, Smith, 182–183
Trans-Alaska pipeline system (TAPS),
  75–78, 87
  Components, 75–76
  Owners, 75

Taxes & dividends, 76
Unfulfilled congressional requirements,
  77–78

U.S. Federal Policy on Research Misconduct,
  128–129
  Fabrication, 129
  Falsification, 129
  Plagiarism, 129

Year 2000 Readiness and Responsibility
  Act, 114–115
Y2K
  Aftermath, 112–113
  Day one preparation, 112
  December 31, 1999, 109
  Millenium bug, 109–110
  Special committee, 110–112
      Assessment, February 1999, 110–111
      Assessment, September 1999, 111–112
      Formation, 110
  Warning, early, 116

Warnings, early
  Bell Laboratories scientific fraud, 130
  Bjork-Shiley heart valve defect, 107
  Challenger space shuttle explosion, 71–73
  Columbia space shuttle explosion, 151
  Exxon Valdez oil spill, 85–86
  Ford Explorer rollover, 140–142
  Ford Pinto recall, 49–50
  Guidant Ancure endograft system, 160
  Indian Ocean tsunami, 182–183
  Kansas City Hyatt Regency skywalk
      collapse, 59
  Northeast blackout, 172–173
  San Francisco-Oakland bay
      bridge earthquake collapse, 96–97
  Y2K software conversion, 116
Watkins, Sherron, 25–26, 34
Williams, Merrill, 32

A Request from the Author

Dear Student,

Most likely, you purchased this textbook because it was required for your class. Once the class is finished, please consider keeping this text rather than selling it as a used textbook. The return on a used book is small. I believe this text (especially Chapters 1, 2, and 16) is much more valuable as a reference during your first five years in industry. Please think about keeping this book.

Sincerely,
Gail Baura

About the Author

Gail Dawn Baura received a BSEE from Loyola Marymount University in 1984, and an MSEE and MSBME from Drexel University in 1987. She received a PhD in Bioengineering from the University of Washington in 1993. Between these graduate degrees, Dr. Baura worked as a loop transmission systems engineer at AT&T Bell Laboratories. Since graduation, she has served in a variety of research and development positions at IVAC Corporation, Cardiotronics Systems, Alaris Medical Systems, and VitalWave Corporation (now Tensys). Dr. Baura is currently Vice President of Research and Chief Scientist at CardioDynamics. Her textbook, *System Theory and Practical Application of Biomedical Signals* (Wiley-IEEE Press, 2002), is part of the IEEE Series in Biomedical Engineering. She will publish a second system theory textbook in 2006. Dr. Baura is a senior member of IEEE, associate editor of *IEEE Engineering in Medicine and Biology* magazine, and a biomedical engineering evaluator for the Accreditation Board of Engineering Technology. She holds 13 issued and 8 pending U.S. patents. Her research interests are the application of system theory to patient monitoring and insulin metabolism.

Printed and bound by CPI Group (UK) Ltd, Croydon, CR0 4YY

03/10/2024

01040410-0001